ADVANCING DESALINATION

Edited by **Robert Y. Ning**

Advancing Desalination

http://dx.doi.org/10.5772/3368
Edited by Robert Y. Ning

Contributors

Keita Kashima, Masanao Imai, Marek Gryta, Peng Wu, Robert Y. Ning, Gamal Khedr, Petr Mikulasek, Jiří Cuhorka, Tzu-Yang HSIEN, Yu-Ling Liu, Thomas Troyer, Abdelkrim Cheriti, Khaled Sekkoum, Talhi Mohamed Fouzi, Bourmita Younes, Nasser Belboukhari, Boulenouar Nouredine, Mohamed F. Talhi, Taleb Safia, Thirugnanasambantham Arun Kumar

Published by InTech

Janeza Trdine 9, 51000 Rijeka, Croatia

Publishing Process Manager Sandra Bakic
Typesetting InTech DTP Team
Cover InTech Design Team

Additional hard copies can be obtained from orders@intechopen.com

Advancing Desalination, Edited by Robert Y. Ning
p. cm.
ISBN 978-953-51-0704-0

For EU product safety concerns:
IN TECH d.o.o., Prolaz Marije Krucifikse Kozulić 3, 51000 Rijeka, Croatia,
info@intechopen.com or visit our website at intechopen.com.

Meet the editor

Dr Robert Y. Ning is Vice President, Science and Business Development of King Lee Technologies based in San Diego, CA, USA. This company was founded in 1977, and is specialized in the chemistry and chemicals used in reverse osmosis (RO) membrane plants. Since 1995, Dr Ning is responsible for science and product development, as well as establishment of a network of process design engineers and service providers around the world to design and service RO plant processes. He has a B.S. degree in Chemistry from Rochester Institute of Technology, Ph.D. in Organic Chemistry from University of Illinois, MBA from Fairleigh Dickinson University, and was a post-doctoral research fellow at California Institute of Technology. He has 25 years of process chemistry experience in the pharmaceutical and biotechnology industries, prior to his current specialization in water and wastewater treatments.

Contents

Preface

This book is a companion volume to two published in 2011 by INTECH titled "Desalination, Trends and Technologies" and "Expanding Issues in Desalination". The term "desalination" used in this series is in the broadest sense of the removal of dissolved, suspended, visible and invisible impurities in seawater, brackish water and wastewater. The purpose of desalination is to make water drinkable, or pure enough for industrial applications like in the processes for the production of steam, power, pharmaceuticals and microelectronics, or simply for attaining acceptable qualities for discharge back into the environment.

This volume touches on Membranes and Systems, Solar Desalination, Reverse Osmosis Process Chemistry and Control, Drinking Water Quality, and Selective Waste Product Removal.

The value of these volumes on the vast topic of desalination is to present the landscape to students, teachers and practitioners, with key concepts and keywords useful in gathering publications through internet search engines. The technologies of desalination of water are advancing as rapidly as the cry of human kind for more availability of quality water supply while minimizing environmental pollution. Contributions to the knowledge-base of desalination are expected to continue to grow exponentially in the coming years.

Robert Y. Ning, Ph.D.
Vice President, Science and Business Development
King Lee Technologies
U.S.A.

Section 1

Membranes and Systems

Chapter 1

Advanced Membrane Material from Marine Biological Polymer and Sensitive Molecular-Size Recognition for Promising Separation Technology

Keita Kashima and Masanao Imai

Additional information is available at the end of the chapter

http://dx.doi.org/10.5772/50734

1. Introduction

Membranes from biological polymers are anticipated practical application as biocompatible materials in separation technology. Biological polymers produced from bioresources are expected to be environmentally compatible polymers and to have great potential as alternatives to various artificial polymers produced from petroleum.

The application of membrane separation in the food industry, medical devices, and water treatment has attracted the attention of biochemical engineering. Membrane separation processes effectively reduce energy cost and CO_2 production. In addition, interest in using natural materials has increased, due to their biocompatibility and their lack of environment load upon disposal. Biopolymer membranes made of cellulose, [1-2], gelatin [3], and chitosan [4] have been anticipated for application in biocompatible separation processing.

1.1. Membrane Desalination

Desalination technology grows exponentially to support water supply from sea water. Today, three billion people around the world have no access to clean drinking water. By 2020, there will be a worldwide 17% short of fresh water needed to sustain the world population. Moreover, 1.76 billion people live in areas already facing a high degree of water stress [5-6].

Generally, desalination can be categorized into two major types: (1) phase-change/thermal process and (2) membrane-based process. Examples of the phase-change process include multi-stage flash, multiple-effect boiling, vapor compression, freezing humidification/dehumidification, and solar stills. Membrane-based processes include reverse osmosis (RO), nanofiltration (NF), ultrafiltration (UF), membrane distillation (MD), and electrodialysis (ED) [7]. Membrane separation technology for desalination is expected to reduce energy consumption.

1.1.1. Artificial Polymer Membrane for Desalination

Previous studies on membrane desalination are listed in Table 1. Various artificial polymers exhibited excellent capability in separation engineering and practical application for desalination, dialysis, and water treatment [8-10].

Authors	Year	Material	Desalination method	Rf.
Hsu, S. T. et al.	2002	PTFE	MD	8
Haddad, R. et al.	2004	Cellulose	NF	18
Peng, P. et al.	2005	PVA/PEG	MD	9
Gazagnes, L. et al.	2007	Ceramic	MD	10
Miao, J. et al.	2008	Chitosan Polysulfone	NF	16
Padaki, M. et al.	2011	Chitosan Polypropylene	NF	17
Zhang, S. et al.	2011	Cellulose	FO	19
Papageorgiou, S. K. et al.	2012	Alginate	Photocatalytic UF	14

MD: Membrane distillation, NF: Nanofiltration, FO: Forward osmosis, UF: Ultrafiltration
PTFE: Polytetrafluoro ethylene, PVA: Polyvinyl alchohol, PEG: Polyethylene glycol

Table 1. Various membranes for desalination.

1.1.2. Biological Polymer Membrane

Alginate is a typical marine biopolymer used as a fouling model in the desalination field [11-12]. Recently, the high performance of desalination of the alginate membrane has been expected to provide highly efficient desalination because sensitive molecular screening characteristics of the alginate membrane have been demonstrated [13]. In addition, alginate-based materials have been developed as support for photocatalysts. Papageorgiou et al. pioneered a hybrid photocatalytic/ultrafiltration process for treating water containing toxic organic compounds [14].

Chitosan has often been investigated for application in desalinating marine biological polymers. Chitosan membrane has strong antibacterial activity in a higher deacetylation degree

http://dx.doi.org/10.5772/50734

[15]. N,O-carboxymethyl chitosan and polysulfone composite membrane cross-linked with epichlorohydrin was recently developed [16].

At 293K and 0.40 MPa, the membrane rejected 90.4% of the Na_2SO_4 solution (1000mg L^{-1}) while the permeate flux was 7.9 kg m^{-2} h^{-1} (Na_2SO_4). In contrast, the membrane rejected 27.4% of the NaCl solution while the permeate flux was 10.8 kg m^{-2} h^{-1} (NaCl). Polypropylene supported chitosan NF-membrane has also demonstrated good desalination ability in acidic pH [17].

Haddad et al. indicated that cellulose acetate nanofiltration (NF) could be adapted to desalination processes [18]. Cellulose ester membrane was also investigated in forward osmosis (FO) for desalination [19]. Forward osmosis has been applied worldwide in recent years as a novel alternative desalination technology for producing fresh water [20].

2. Alginates

Alginic acid is abundantly produced by marine biological resources, especially brown seaweed. The first description of alginate as a preparation of "algic acid" from brown algae was provided and demonstrated by British chemist E. C. C. Stanford, with a patent dated 12 January 1881 [21]. In 1896, A. Krefting successfully prepared a pure alginic acid. Kelco Company began commercial production of alginates in 1929 and introduced milk-soluble alginic acid as an ice cream stabilizer in 1934 [22].

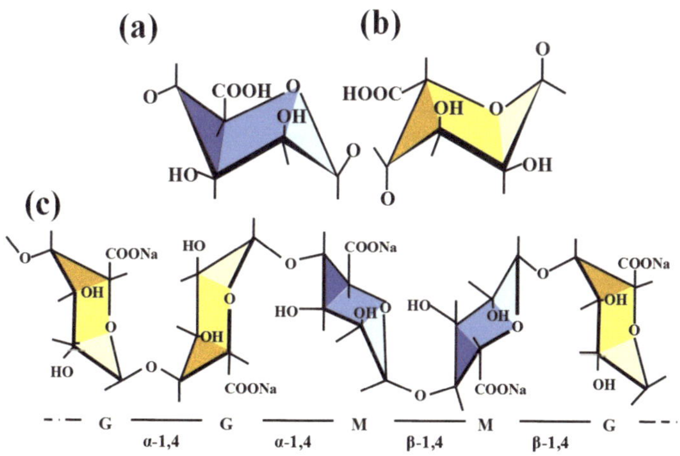

Figure 1. Alginate composition. (a) β-D-mannuronic acid. (b) α-L-guluronic acid. (c) Structural formula of sodium alginate molecule.

Alginates have been conventionally applied in the food industry as thickeners, suspending agents, emulsion stabilizers, gelling agents, and film-forming agents [23].

Sodium alginate is a typical hydrophilic polysaccharide. It consists of a linear copolymer composed of two monomeric units, 1,4-linked β-D-mannuronic acid (Figure 1a) and α-L-guluronic acid (Figure 1b), in varying proportions. These two uronic acids have only minor differences in structure, and they adopt different chair conformations such that the bulky carboxyl group is in the energetically favored equatorial position [24].

The physical properties (e.g., viscosity and mean molecular weight) of sodium alginate are very susceptible to physicochemical factors (e.g., pH and total ionic strength). At near-neutral pH, the high negative charge of sodium alginates due to deprotonated carboxylic functional groups induces repulsive inter- and intra-molecular electrostatic forces. The change of ionic strength in a sodium alginate aqueous solution has a significant effect, especially on the polymer chain extension [25-27].

2.1. Chemical Formation

An alginate molecular chain was constructed using three types of polymeric block: homopolymeric blocks of mannuronic acid (M-M), guluronic acid (G-G), and blocks with an alternating sequence in varying proportions (M-G) [28] (Figure 1c).

The M-M block consists of 1e→4e linked β-D-mannuronic acid chains with the monosaccharide units in a $^4C_1$ chair conformation. Regions in which β-D-mannuronic acid predominates have been predicted to form an extended ribbon structure, analogous to cellulose [29].

The G-G block is composed of 1→4 diaxially linked α-L-guluronic acid residues in a $^1C_4$ chair conformation. It forms a buckled chain [30]. The molecular construction of the G-G block has been confirmed experimentally by X-ray diffraction analysis of the partial hydrolysis products of alginate. The mass fraction of these blocks is basically derived from a natural species of brown algae. At present, the production of new tailor-made alginates has been prompted by the availability of C-5 epimerases, which facilitate extremely efficient tuning of both composition and physicochemical properties of the polysaccharide. In particular, the epimerase AlgE4, which enables the conversion of M-M blocks into alternating sequences in a processive mode of action [31], has provided new alginates with interesting properties. In this respect, besides the remarkable increase in syneresis displayed by the AlgE4 treated samples, a much higher stability of the gel is directly correlated with the presence of long alternating sequences [32-33].

2.2. Gelling Ability

Sodium alginate rapidly forms a gel structure with the presence of divalent cations such as Ca^{2+}, resulting in a highly compacted gel network [34]. Spherical gel particles of calcium alginate are often investigated and applied as a carrier of immobilized enzyme [35], a drug delivery capsule [36], a carrier of entrapped living cells [37-38], and a food supplement [39]. However, the formation of the alginate membrane has not been investigated as much.

The basic ability of alginate to gel is related to its specific ion-binding characteristics [40]. The variation in gel strength has been analyzed in terms of modes of binding of cation by the various block structures that occur within the alginate molecular chain. Experiments involving equilibrium dialysis of alginate have demonstrated that the selective binding of certain alkaline earth metal ions (e.g., strong and cooperative binding of Ca relative to Mg) increases markedly with increasing content of G-G block in the chains. M-M block and M-G block had almost no selectivity [41]. Regions of homopolymeric blocks of α-L-guluronic acid chelate the alkaline earth metal ions because of the spatial arrangement of the ring and hydroxyl oxygen atoms, and thus create a much stronger interaction [24]. These homopolymeric blocks of α-L-guluronic acid junction zones are constructed mainly of a cross-linked area called an "Egg-box," where the Ca^{2+} ions are located as the "Egg" components [42] (Figure 2). NMR studies of lanthanide complexes of related compounds suggested a possible binding site for Ca^{2+} ions in a single alginate chain [43].

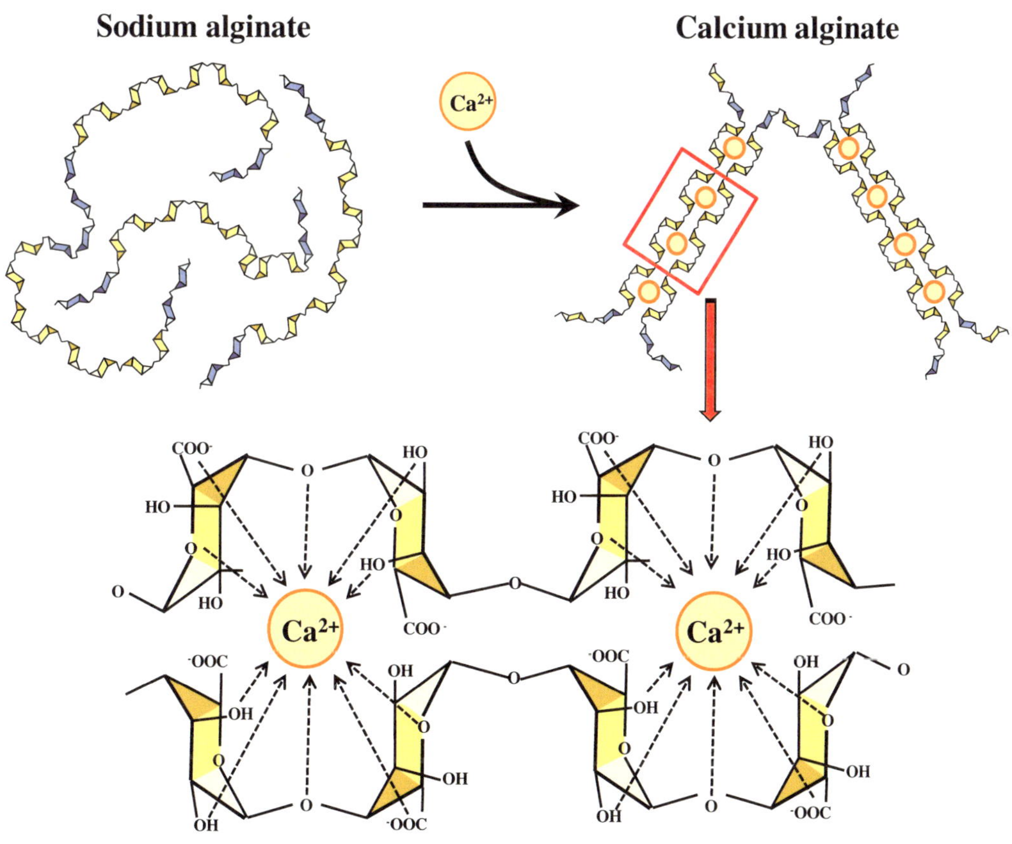

Figure 2. Gelation of homopolymeric blocks of α-L-guluronic acid junction with calcium ions. Binding of divalent cations by alginate: the "Egg-box" model.

3. Membrane Preparation

Many kinds of biopolymer membrane have been utilized and developed in food and biological applications. In general, biopolymer membrane was prepared by casting (e.g., cellulose acetate) [44] and chitosan [45]). Spherical gel particles of sodium alginate have often been investigated and applied. However, the formation of alginate membrane has been less investigated. This section provides a general description of the preparation of various alginate membranes.

3.1. Previous Studies on Membrane Preparation

In recent years, alginate membranes have been investigated in diverse ways (e.g., pervaporation, immobilized cell reactor, and ultrafiltration). Previous studies on alginate membrane are listed in Table 2. Teixeira et al. prepared yeast-cell-occupied calcium alginate membrane [46]. Zhang and Franco prepared a calcium alginate membrane for measuring effective diffusivities using the diffusion-cell technique [47]. Grassi et al. determined the drug diffusion coefficient in a calcium alginate membrane [48]. Alginate membrane prepared by low concentration cross-linker needed support matrix (e.g. glass fiber filter) to maintain flat membrane [49].

3.1.1. Cross-linker

Calcium chloride is basically used as a cross-linker in many investigations of calcium alginate membranes. Calcium sulphate and calcium acetate have also been used as cross-linkers for calcium alginate membrane preparation [23, 50-51]. Other cations (Ba^{2+}, Zn^{2+}) have been used as cross-linkers for preparing alginate membrane [49, 51-52]. Barium chloride provided more improved stability than calcium chloride [49]. Zinc acetate can cause denser cross-linking and less selectively than calcium used with sodium alginate [51].

Sodium alginate membrane cross-linked by glutaraldehyde was applied for acetic acid separation from acetic acid aqueous solution. The membrane was also applied for separating isopropanol from its aqueous solution [53]. Experimental evidence from IR spectroscopy, wide angle X-ray diffractometry, and swelling measurements enabled characterization of the reaction between sodium alginate and glutaraldehyde. The aldehyde groups increased with increasing glutaraldehyde content in the reaction solution [54].

Kalyani et al. prepared a sodium alginate membrane with phosphoric acid for separating ethanol aqueous solution. Phosphoric acid established a linkage with sodium alginate through ester formation, as confirmed by FTIR [55].

http://dx.doi.org/10.5772/50734

Table 2-1. Previous investigation results of alginate membrane (1).

Authors	Year	Base cain	Concentration of base chain	Cross-linker	Concentration of cross-linker	Investigation	Tested material	Comments	Rf.
Hubble, J. et al.	1985	Sodium alginate	2-4% w/v	Calcium chloride Barium chloride	0.05M 0.05M	Ultrafiltration	Bovine serum albumin Concanavalin A Ferritin	Membrane was supported by glass fibre filter.	49
Andreopoulos, A. G. et al.	1987	Sodium alginate	1.8% w/w	Calcium sulphate dihydrate	Unknown	Vapour sorption	Methyl methacrylate	-	50
Julian, T. N. et al.	1988	Sodium alginate	3% w/w	Calcium acetate	0.1-1.0M	Permeability	Acetaminophen	-	23
Teixeira, J. A. et al.	1994	Sodium alginate	3% w/v	Calcium chloride	2% w/v	Diffusion coefficient CO_2 evolution	Glucose Malic acid	Yeast cell was immobilized in the membrane.	46
Aslani, P. and R. A. Kennedy	1996	Sodium alginate	3% w/w	Calcium acetate Zinc acetate	0.1-0.7M 0.1-0.7M	Diffusion coefficient	Acetaminohen	Drug diffusion	51
Yeom, C. K., and Lee, K.-H.	1998	Sodium alginate	2.5% w/w	Glutaraldehyde	0-20% v/v	Pervaporation	Ethanol	-	54
Zhang, W. et al.	1999	Sodium alginate	2 % w/v	Calcium chloride	2% w/v	Diffusion coefficient	Glucose Lactic acid	*Lactobacillus rhamnosus* was immobilized.	47
Yang, G. et al.	2000	Sodium alginate blend with Cellulose	0 - 8% w/w 8 - 0% w/w	Calcium Chloride	5% w/w	Pervaporation	Ethanol	-	56
Wang, X. P.	2000	Sodium alginate coated on Polyacrylonitrile	1% Commercial membrane	1,6-Hexanediamine or Poly(vinyl alcohol)	0.25% 1%	Pervaporation	Acetic acid	-	59
Grassi, M. et al.	2001	Sodium alginate	Unknown	Calcium chloride	0.05 M	Diffusion coefficient	Theophylline	Drug diffusion	41
Toti, U. S. et al.	2004	Sodium alginate	5% w/v	Glutaraldehyde	1% v/v	Pervaporation	Acetic acid Isopropanol	Different viscosity grade sodium alginatew were tested.	53
Kanti, P. et al.	2004	Sodium alginate blend with Chitosan	3% w/w 3% w/w	Glutaraldehyde	5% v/v	Pervaporation	Ethanol	-	60

Table 2-2. Previous investigation results of alginate membrane (2).

Authors	Year	Base Cain	Concentration of base chain	Cross-linker	Concentration of cross-linker	Investigation	Tested material	Comments	Rf.
Rhim, J.-W.	2004	Sodium alginate	2% w/v	Calcium chloride	0.04 - 0.12 g / 4g alginate 10 - 50 w/v	Physical and Mehanical properties	-	Two different methods of cross-linking were tested.	69
Smitha, B. et al.	2005	Sodium alginate blended with Chitosan	3% w/w 3% w/w	none	-	Direct methanol fuel cell	Methanol	Cross-link was used only polyion comples.	62
Zimmermann, H. et al.	2007	Sodium alginate	0.7% w/v	Barium chloride	20mM	Physical and Biological properties		NMR, CLSM, AFM, Burst pressure, Water flow	52
Kalyani, S. et al.	2008	Sodium alginate	3% w/w	Phosphoric acid	3.5 vol.%	Pervaporation	Ethanol	-	55
Reddy, A. S. et al.	2008	Sodium alginate blend with Chitosan	2% w/w 2% w/w	Calcium chloride and Maleic anhydride	2% 3.5 % w/w	Pervaporation	1,4-Dioxane	-	61
Kashima, K. et al.	2010	Sodium alginate	10g/L	Calcium chloride	0.05-1.0M	Molecular size Mechanical property Water permeability	Urea Glucose Methyl Orange Indigo Carmine Bordeaux S	Superior molecular size screening ability was found.	63
Saraswathi, M. et al.	2011	Sodium alginate blend with dextrin	20-0% w/v 0-20% w/v	Glutaraldehyde	Unknown	Pervaporation	Isopropanol	-	58
Chen, J. H., et al.	2012	Sodium alginate Hydroxyl ethyl cellulose	2.2% w/w 0.19% w/w	Glutaraldehyde	0.5% w/w	Adsorption	Cd (II) ion	-	57
Papageorgiou, S. K., et al.	2012	Sodium alginate	10.7% w/w	Calcium chloride	10%	Photocatalytic UF	Methyl orange	Alginate fiber stabilized TiO_2	14

3.1.2. Hybrid membrane with other polymers

Many efforts have been made to increase the performance of the alginate membrane by blending it with different hydrophilic polymers. Alginate-cellulose using a calcium ion cross-link was investigated in the permeation flux of ethanol aqueous solution for pervaporation [56]. A novel porous composite membrane was prepared using sodium alginate and hydroxyl ethyl cellulose hybrid as an immobilization matrix for humic acid, then cross-linked by glutaraldehyde [57]. Hybrid membranes of sodium alginate and dextrin were prepared by casting followed by cross-linking with glutaraldehyde and used for pervaporation separation of isopropanol aqueous solution [58]. Casting an aqueous solution of alginate with 1,6-hexanediamine or poly (vinyl alcohol) on a hydrolyzed microporous polyacrylonitrile membrane was characterized by pervaporation separation of acetic acid aqueous solution [59].

The most employed alginate hybrid material was chitosan. Polymer complex membranes made by blending 84% deacetylated chitosan and sodium alginate followed by cross-linking with glutaraldehyde were tested for separating ethanol from ethanol aqueous solution [60]. Sodium alginate and chitosan hybrid membranes were cross-linked with maleic anhydride for separating 1,4-dioxane aqueous solution. Such a membrane has good potential for breaking the aqueous azeotrope 1,4-dioxane [61].

An alginate-chitosan membrane without a cross-linker could be prepared practically. The structural formation of a chitosan-alginate ion complex was attained between the anion group ($-COO^-$) of sodium alginate and the protonated cation group ($-NH_3C$) of chitosan [62].

3.2. Preparation of a Flat Alginate Membrane

Our original procedure to prepare calcium alginate membrane by casting was as follows. One gram sodium alginate was dissolved in 100mL water. Sodium alginate samples were provided by Wako Pure Chemical Industries, Ltd. (Osaka, Japan) and KIMICA Corporation (Tokyo, Japan). Calcium chloride (0.05M to 1.0M) was also dissolved in water. Twenty grams of the sodium alginate solution was dispensed on a Petri dish and then completely dried in desiccators at room temperature (298K) for one week. A dried thin film of sodium alginate appeared on the Petri dish. Next, calcium chloride aqueous solution was added directly to the dried thin film of sodium alginate in the Petri dish. A calcium alginate membrane quickly formed in the Petri dish at room temperature. After 20min, the swollen membrane was separated from the Petri dish and then left in the dish for an additional 20min. The membrane was immersed for a total of 40min in the calcium chloride aqueous solution. The formed calcium alginate membrane was soaked in pure water to remove excess calcium chloride aqueous solution, then stored in pure water [63].

The fundamental gelling mechanism of alginate polymer was the ionic binding reaction between G-G blocks and divalent cations, such as Ca^{2+}. Alginate has high potential of ion exchange. Cross-linking quickly started in the alginate solution. A calcium alginate gel particle was easily obtained by injecting the sodium alginate solution into the calcium aqueous solution [64]. In contrast, the quickly gelling reaction inhibited the preparation of a flat alginate membrane. To overcome rapid gelling, sodium alginate aqueous solution was first dried, and then the cross-linker aqueous solution was directly introduced into the dried alginate surface. As a result, a calcium alginate membrane having a flat surface was successfully prepared. The advanced feature of flat alginate membrane preparation was originally examined by Kashima et al. [63].

3.3. Evaluation of Components in the Alginate Polymer Chain

As mentioned in chapter 2, components in the alginate polymer chain important factors in investigating the properties of the alginate gel membrane. Two uronic acids, β-D-mannuronic acid (M) and α-L-guluronic acid (G), were constituents of the alginate molecular chain. The homopolymeric blocks of α-L-guluronic (G-G block) in the alginate chain are constructed mainly of a cross-linked zone. Hence, G-G blocks perform a dominant role in the mechanical strength and the mass-transfer characteristics of the calcium alginate membrane [65].

3.3.1. Qualitative analysis of uronic acid

Mannuronic acid lactone was used as the standard component of uronic acid. As the standard solution, various concentrations of mannuronic acid lactone were dissolved in water. The concentrations were determined by Bitter-Muir's carbazole sulfuric acid method [66], and the concentration of the colored solution was measured by optical density at 530nm. The analysis produced good intensity and accuracy of coloration [67].

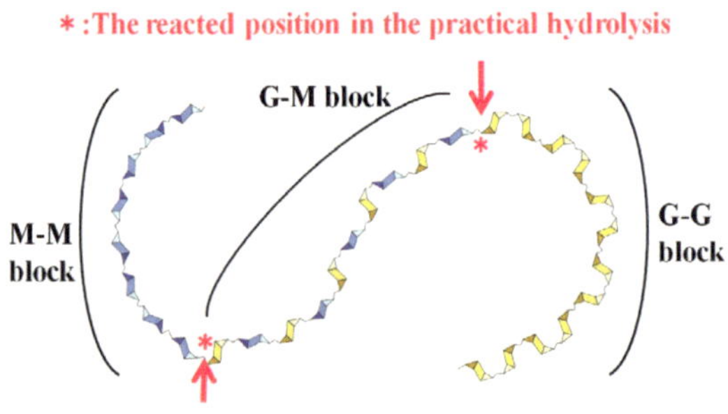

Figure 3. Sodium alginate molecular chain. The hydrolyzed site is indicated by an asterisk (*). Sodium alginate was then separated into three molecular chain blocks: M-G, M-M, and G-G.

3.3.2. Partial hydrolysis of the alginate molecular chain to determine the mass fraction of homopolymeric blocks

The mass of uronic acid in the actual alginate chain was determined by partial hydrolysis combined with Bitter-Muir's carbazole sulfuric acid method. Partial hydrolysis protocols were employed according to a previous method [67]. Figure 3 illustrates the alginate molecular chain and homopolymeric blocks. The reacted position of partial hydrolysis is marked by an asterisk (*). The sodium alginate chain was separated into three molecular chain blocks (M-G, M-M, and G-G).

http://dx.doi.org/10.5772/50734

Sodium alginate (0.5g) was dissolved in 0.3M HCl (50mL). The resulting solution was heated in an electrical blast-drying chest (373K) for 2h to promote partial hydrolysis. The partial hydrolysis solution was then centrifuged ($3000 min^{-1}$, 15min), and a sample solution of the M-G block was obtained as the supernatant.

The precipitate was mixed with pure water (10mL), and 3M NaOH was added to aid dissolution. The concentration was then adjusted to 1% by the addition of pure water, and NaCl was introduced to achieve 0.1M of sodium alginate. The solution was adjusted to pH 2.9 using 2.5M HCl and then centrifuged ($3000 min^{-1}$, 15min). The sample solution of the M-M block was obtained as the supernatant.

After filtration, the precipitate was mixed with pure water (10mL) and dissolved by adding 3M NaOH, yielding the sample solution of the G-G block. As a result, three sample solutions (M-G, M-M, and G-G) were obtained.

3.3.3. Mass fraction of homopolymeric blocks of α-L-gluronic acid

The mass of the M-G block in the sodium alginate (W_{MG}) was directly obtained from the concentration of the M-G block sample. The mass of the M-M block (W_{MM}) and that of the G-G block (W_{GG}) in the sodium alginate were also obtained independently in the same manner. The mass fraction of α-L-guluronic acid in the sodium alginate (F_G) was then calculated using the following formula:

$$F_G = \frac{W_{GG} + W_{MG} \times P}{W_{GG} + W_{MG} + W_{MM}} \tag{1}$$

where P is the partial mass fraction of α-L-guluronic acid in the M-G block. The polymeric structure of the calcium alginate gels was constructed mainly by intermolecular ionic bonds in the homopolymeric blocks of the α-L-guluronic acid junction zone, in combination with Ca^{2+} [42]. Therefore, in our study, P is assumed to be negligible (P = 0), and Eq. (1) is rearranged as follows.

$$F_{GG} = \frac{W_{GG}}{W_{GG} + W_{MG} + W_{MM}} \tag{2}$$

The mass fraction of the homopolymeric blocks of α-L-guluronic acid (F_{GG}) was therefore obtained from natural resources. It was considered a key factor in regulating membrane properties.

3.4. Morphology of the Calcium Alginate Membrane

No stable calcium alginate membrane was obtained using a very low concentration (e.g. less than 0.01M) of $CaCl_2$. It looked like jelly (Figure 4a). A stable, flat, thin membrane was obtained with more than 0.05M $CaCl_2$ solution as a cross-linker (Figure 4b). The membrane became transparent with increasing $CaCl_2$ concentration (Figures 4c, d).

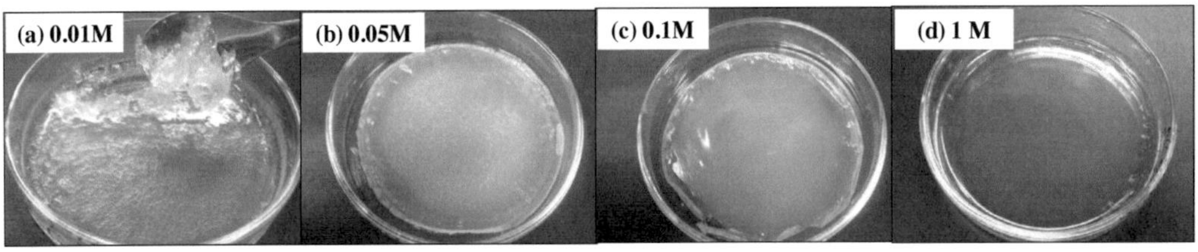

Figure 4. Pictures of calcium alginate membrane. (a) $CaCl_2$: 0.01M, (b) $CaCl_2$: 0.05M, (c) $CaCl_2$: 0.1M, (d) $CaCl_2$: 1M.

In scanning electron microscopy (SEM) (Miniscope TM-1000, Hitachi, Ltd., Tokyo, Japan), the surface of the membrane appeared to be smooth. No pores were observed on the surface. The higher F_{GG} membrane had a dense surface (Figure 5a), whereas the lower F_{GG} membrane appeared harsh (Figure 5b). Regardless of F_{GG}, the membrane surface observed by SEM became smoother with increasing $CaCl_2$ concentration (Figures 5a, b, c, d).

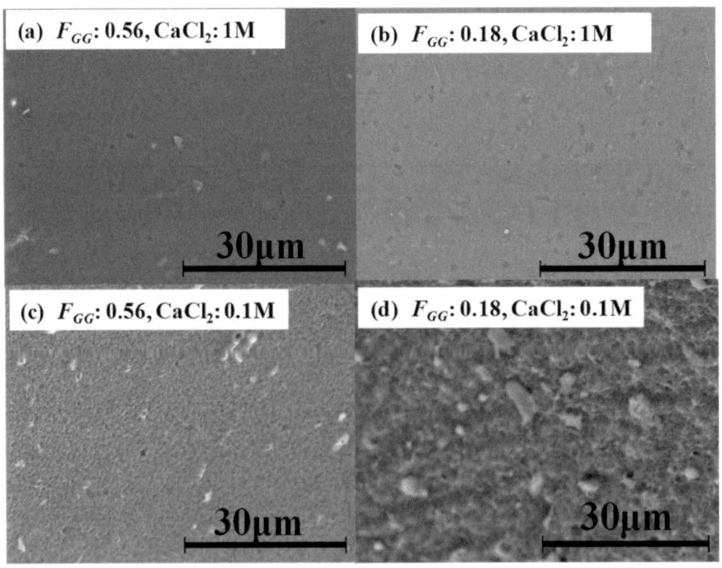

Figure 5. Scanning electron microscopy (SEM) images of the surface of the calcium alginate membrane. (a) F_{GG}: 0.56, $CaCl_2$: 1M. (b) F_{GG}: 0.18, $CaCl_2$: 1M. (c) F_{GG}: 0.56, $CaCl_2$: 0.1M. (d) F_{GG}: 0.18, $CaCl_2$: 0.1M.

Figure 6 presents scanning probe microscope (SPM) (S-image SII Nano Technology, Inc., Tokyo, Japan) images of the membrane surfaces. SPM can determine the morphology of the membrane surface by using the physical force (e.g., atomic force) between the cantilever and the sample membrane. In our case, dynamic force mode/microscopy (DFM) was used for observation. DFM is a measurement technique based on making the cantilever resonant to detect gravitation and repulsive forces against the sample surface. It is a morphology measurement mode for stable observation of relatively sticky, uneven, and soft samples (e.g., biopolymers). The distribution of membrane asperity clearly decayed with increasing F_{GG} (Figures 6a, b). In contrast, with a low concentration of calcium, the distribution of membrane asperity changed little (Figures 6c, d). These results suggest that the molecular framework was condensed by increasing F_{GG}. With higher $CaCl_2$ concentration (1.0M), the effect of F_{GG} became dominant and made a smooth surface. However, with lower $CaCl_2$ concentration (0.1M), the effect of F_{GG} was insignificant, and the membrane surface did not become smooth.

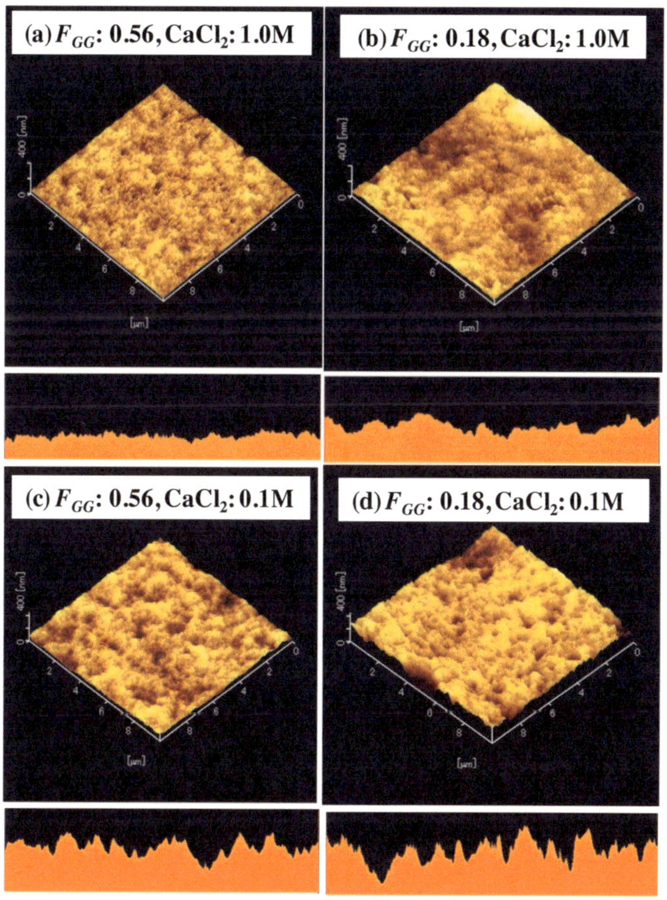

Figure 6. Scanning probe microscopy (SPM) views of the surface of the calcium alginate membrane. (a) F_{GG}: 0.56, $CaCl_2$: 1M. (b) F_{GG}: 0.18, $CaCl_2$: 1M. (c) F_{GG}: 0.56, $CaCl_2$: 0.1M. (d) F_{GG}: 0.18, $CaCl_2$: 0.1M.

4. Mechanical Properties of the Calcium Alginate Membrane

Investigation of mechanical properties is important for practical application. The following section describes the maximum stress and strain at membrane rupture of the calcium alginate membrane involved with calcium concentration and F_{GG}.

4.1. Investigation of Mechanical Strength

The maximum stress and strain at membrane rupture is evaluated by rheometer as a general test of the mechanical properties of the polymer membrane. A swollen membrane sample (10mm wide and 30mm long) was mounted in the rheometer (CR-DX500, Sun Scientific Co., Ltd., Tokyo, Japan) with a crosshead speed of 2mm/s. Maximum stress [N/m^2] at membrane rupture was evaluated based on the loading force divided by the cross-sectional area of the membrane. Maximum strain was evaluated as the percentage by which the length increased at membrane rupture divided by the original length of the membrane sample. The relationship between maximum stress and strain with deacetylation degree was investigated as an elastic property of the chitosan membrane using this method [4].

4.2. Effect of Calcium Concentration

Figure 7 depicts the effect of $CaCl_2$ concentration on the maximum stress at membrane rupture. The maximum stress increased with increasing $CaCl_2$ concentration as a cross-linker. With higher F_{GG} (F_{GG} = 0.56), the maximum stress increased remarkably, especially with higher $CaCl_2$ concentration.

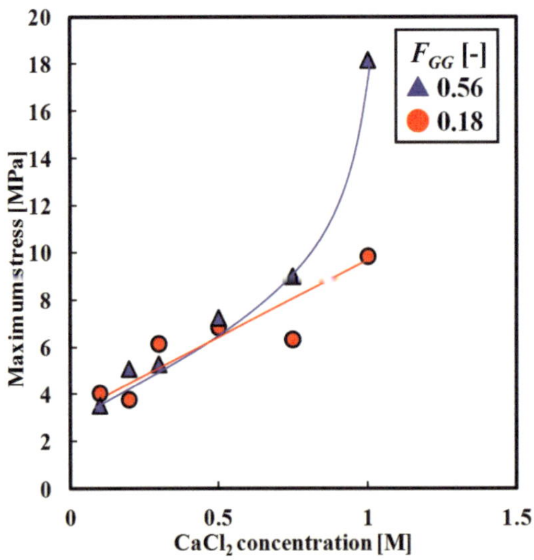

Figure 7. Effect of $CaCl_2$ concentration on maximum stress at membrane rupture.

http://dx.doi.org/10.5772/50734

In contrast, the maximum strain at membrane rupture was remarkably reduced by adding $CaCl_2$ (Figure 8). The cross-linked site became more highly populated with increasing $CaCl_2$ concentration. It was resulted increasing the mechanical strength [63]. Using $CaCO_3$ as a cross-linker, the mechanical properties exhibited profiles similar to those using $CaCl_2$ [68].

4.3. Effect of Mass Fraction of the Homopolymeric Blocks of α-L-Guluronic Acid (F_{GG})

Mechanical strength and elastic characteristics apparently changed with F_{GG}. Maximum stress increased remarkably with increasing F_{GG} at the same Ca^{2+} concentration (Figure 7). In contrast, when the membrane ruptured, maximum strain was remarkably reduced with increasing F_{GG} (Figure 8). Increasing F_{GG} obviously enhanced the polymeric framework of the membrane.

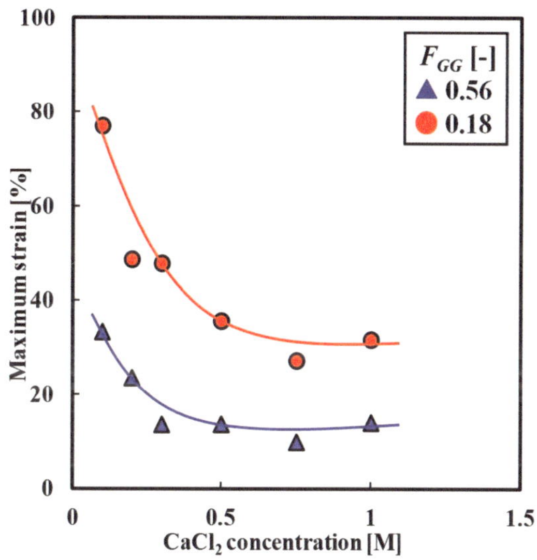

Figure 8. Effect of $CaCl_2$ concentration on maximum strain at membrane rupture.

4.4. Effect of Cross-Linking Methods

The mechanical properties of calcium alginate membranes prepared from two different $CaCl_2$ treatments were examined by Rhim [69]. One is the direct mixing of $CaCl_2$ into a membrane-making solution (mixing membrane). The other is the immersion of alginate membrane into $CaCl_2$ solution (immersion membrane). With the mixing method, maximum stress and maximum strain at the break of the mixing membrane did not change with increased addition of $CaCl_2$. In contrast, for the immersion membrane, the maximum stress increased and the maximum strain decreased with increased addition of $CaCl_2$. The membrane became rigid. In the immersion method, the mechanical characteristics were strongly influenced by $CaCl_2$ concentration.

4.5. Effect of Relative Humidity

The effect of relative humidity on the mechanical properties of the calcium alginate membrane was examined at relative humidities of 59%, 76%, 85% and 98% at room temperature for 8 days [70]. As relative humidity increased, maximum strain increased and maximum strength decreased.

4.6. Comparison with Other Membranes

Figure 9 indicates the mechanical properties of various polymer membranes. The calcium alginate membrane exhibited high stress and low strain at rupture. It had better mechanical properties than other biopolymer membranes (e.g., chitosan [4] and cellulose acetate [71]). The higher F_{GG} membrane had higher mechanical strength at rupture, with elasticity. However, the lower F_{GG} membrane was flexible and had desirable mechanical strength.

For comparison, the polytetrafluoroethylene (PTFE)/polyvinyl alcohol (PVA) composite membrane had stronger mechanical strength and very low maximum strain, with rigidity [72] (Figure 9).

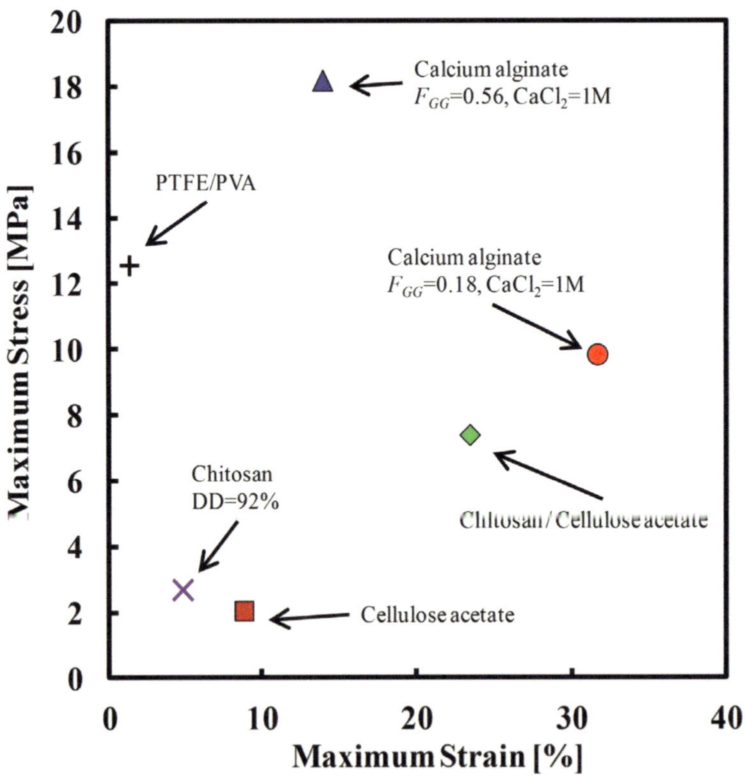

Figure 9. Mechanical strength of alginate membrane compared with various polymer membranes.

5. Water Content in a Swollen Membrane

The water content of a hydrophilic membrane influences the diffusion phenomena and water permeability. In general, polymer membranes having higher water content have higher water permeability, that has been reported in cellulose acetate membranes [73]. As water occupied mainly the void of the membrane, volumetric water content is often regarded as the void fraction of the membrane structure [74].

5.1. Evaluation of Water Content

The volumetric water content of the swollen membrane was not measured directly. Instead, it was evaluated from the mass-based water content (H_M) using gravimetric methods. The swollen membrane is assumed to have equilibrium water content. Excess water attached to the membrane surface was removed using filter paper. The mass of the swollen sample (w_e) was measured initially, then, after drying (333K for 24h), the mass of the dried membrane at equilibrium state (w_d) was measured. For strict analysis, w_d included "bonding water" on polymer networks. It is assumed to be negligible in the following description [45].

The difference between w_e and w_d represents the mass of the total contained water (w_t).

$$\mathrm{w_t = w_e - w_d} \tag{3}$$

The mass-based water content of the swollen membrane (H_M) was then calculated using the following equation.

$$\mathrm{H_M = \frac{w_e - w_d}{w_e} = \frac{w_t}{w_e}} \tag{4}$$

The volume of water contained in the membrane void was evaluated from w_t/ρ_W. The apparent volume of the swollen membrane was estimated as w_e/ρ_M. The apparent density of the swollen membrane ρ_M was determined from the mass of the swollen membrane w_e divided by the apparent volume of the swollen membrane, which was calculated from the membrane area (square with 4cm sides) and its thickness. The estimated volumetric water fraction of the swollen membrane (H_V) was calculated using Eq. (5).

$$H_V = \frac{(w_t / \rho_W)}{(w_e / \rho_M)} \tag{5}$$

H_V is often employed as the void fraction (porocity) of the swollen-state membrane. The volumetric water content (H_V) in the calcium alginate membrane is presented in Figure 10. H_V gradually decreased with increasing $CaCl_2$ concentration.

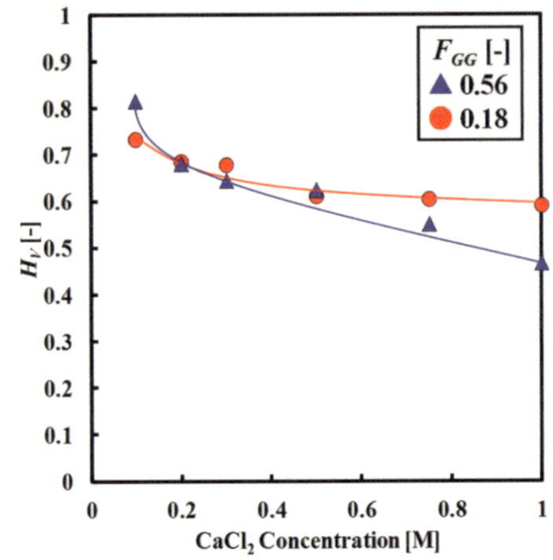

Figure 10. Effect of $CaCl_2$ concentration on volumetric water content H_V.

5.2. Effect of Calcium Ion and F_{GG} on Water Content in Membrane

The dry-based water content of calcium alginate gel beads loading sucrose has been investigated for encapsulation-dehydration of plant germplasm [75]. The dry weight of the beads decreased, and the water content increased with increasing cross-linking time (10 to 30min). The sucrose was diffused to the outer aqueous phase, and then the water penetrated into the gel beads. This is understood as osmotic phenomena surrounding the gel particles. The mass fraction of unfrozen water compared to total water content was also investigated as the thermal property of the gel beads. It increased within 5 to 15min to achieve maximum level (23%), and then declined to minimum level (17%) at 30min.

The H_V of the swollen membrane gradually decreased with increasing F_{GG} [65]. The effect of F_{GG} was especially strong with higher $CaCl_2$ concentration (Figure 10). The lower F_{GG} membrane had higher water content, in spite of the high $CaCl_2$ concentration. The cross-linking molecular chain decreased with lower F_{GG}.

5.3. Stability of the Swollen Membrane

The stability of the swollen membrane is important to long-life use in practical applications. Rhim focused on the gravimetrical change of the membrane before and after practical use. Stability was evaluated by the dry-base mass of the membrane [69]. Rhim focused on the gravimetrical change of the membrane before and after practical use. Stability was evaluated by the dry-base mass of the membrane [69]. The membrane mass decreased 16% to 20% with increasing soaking temperature (298K to 353K) in aqueous phase. This change was induced by dissolving the membrane matter into aqueous phase. In contrast, the change in membrane mass did not present any significant difference with $CaCl_2$ concentration. The stability of the membrane was affected by soaking temperature but did not depend on $CaCl_2$ concentration.

http://dx.doi.org/10.5772/50734

5.4. Void Fraction of Membrane

The water content of a membrane can be regarded as an indicator of the void fraction (ε) of the membrane structure [74]. Al-Rub et al. found that membrane distillation mass flux increased linearly with the membrane void fraction, whereas the temperature difference increased slightly with an increase in membrane void fraction [76]. This is due to the fact that a higher void fraction means that more mass-transfer channels exist for diffusion; hence, higher flux results. The void fraction of commercial microfiltration membranes varies from 60% to 90%, depending on material type, membrane form (flat sheet or hollow fiber), and manufacturing method [77]. The calcium alginate membrane has a high void fraction (50% to 90%) [63].

6. Mass-transfer Characteristics

The diffusivity in calcium alginate "beads" has often been investigated. The effective diffusion coefficient of the alginate "membrane" was originally reported by Kashima et al. [63]. The effective diffusion coefficients are listed in Table 3 and plotted in Figure 11.

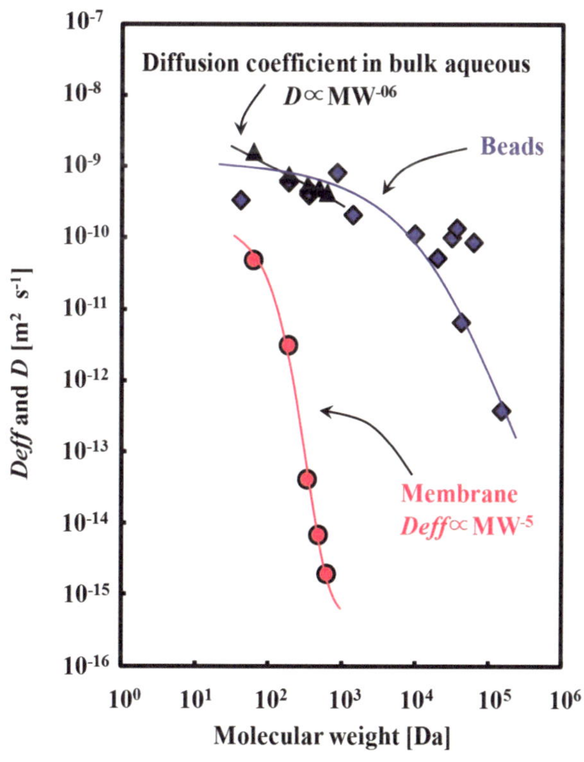

Figure 11. Effective diffusion coefficient of calcium alginate gel.

Table 3. Effective diffusion coefficient in calcium alginate gels.

Authors	Year	Base chain	Concentration of base chain	Cross-linker	Concentration of cross-linker	Gel type	Diffusion conponent	MW [Da]	Effective diffusion coefficient $[m^2 s^{-1}]$	Rf.
Mehmetoglu, U.	1990	Calcium alginate	5%	Calcium chloride	0.05M	Beads	Sucrose	342	4.10×10^{-10}	84
Longo, M. A. et al.	1992	Calcium alginate	2% w/v	Calcium chloride	3% w/v	Beads	*Bacillus* total protease	30000	1.04×10^{-10}	85
							Bacillus metalloprotease	35000	1.40×10^{-10}	
							Serratia total protease	60000	8.90×10^{-11}	
Kikuchi, A. et al.	1999	Calcium alginate	2% w/w	Calcium chloride	0.1M	Beads	Ca^{2+}	40	3.47×10^{-10}	86
							Brilliant Blue	826	8.41×10^{-10}	
							Dextran	9400	1.16×10^{-10}	
								19000	5.39×10^{-11}	
								40500	6.73×10^{-12}	
								145000	3.90×10^{-13}	
Dembczynski, R. and T. Jankowski	2000	Calcium alginate	1% w/v	Calcium chloride	4% w/v	Beads	Glucose	180	6.58×10^{-10}	39
							Fructose	180	6.18×10^{-10}	
							Sucrose	342	4.70×10^{-10}	
							Lactose	342	4.51×10^{-10}	
							Vitamin B12	1355	2.16×10^{-10}	
Koyama, K. and M. Seki	2004	Calcium alginate	2% w/v	Calcium chloride	1% w/v	Beads	Glucose	180	6.50×10^{-10}	87
Kashima, K. et al.	2010	Calcium alginate	1% w/v	Calcium chloride	1M	Membranes	Urea	60	5.06×10^{-11}	65
							Glucose	180	3.22×10^{-12}	
							Methyl Orange	327	4.24×10^{-14}	
							Indigo Carmine	466	7.07×10^{-15}	
							Bordeaux S	604	2.02×10^{-15}	

http://dx.doi.org/10.5772/50734

6.1. Analysis of Mass-Transfer

The typical procedure to measure mass-transfer flux is as follows. The overall mass-transfer coefficient K_{OL} was determined by measuring the mass-transfer flux based on Eqs. (6) and (7). The membrane was sandwiched between twin glass mass-transfer cells that were placed in a thermostatic bath (298K).

$$\ln\left(1 - \frac{2C_s}{C_{fi}}\right) = -2\frac{A}{V}K_{OL}t \tag{6}$$

$$K_{OL}^{-1} = k_{L1}^{-1} + k_m^{-1} + k_{L2}^{-1} \tag{7}$$

Both aqueous phases were sufficiently stirred to create a fully developed turbulent flow. Film mass-transfer resistances k_{L1}^{-1} and k_{L2}^{-1} in the overall mass-transfer resistance K_{OL}^{-1} were ignored under fully turbulent conditions. In this case, K_{OL} did not depend on stirring rate. Therefore, it directly indicated the membrane mass-transfer coefficient km ($km = D_{eff}/l$). The effective diffusion coefficient in the membrane (D_{eff}) was evaluated from km. The initial thickness of the swollen membrane l was measured with a micrometer (Mitutoyo Corporation, Kawasaki, Japan). The molecular-size screening capability of the calcium alginate membrane was investigated by measuring mass-transfer flux using various molecular-size indicators (Urea 60 Da, Glucose 180 Da, Methyl orange 327 Da, Indigo carmin 466 Da, and Bordeaux S 604 Da) [63].

The concentration of the stripping solution was determined by a spectrophotometer (UV Mini 1240, Shimadzu, Kyoto, Japan). The absorbances of the color pigments employed (Methyl orange, Indigo carmine, and Bordeaux S) were measured based on the maximum wavelength (Methyl orange 462nm, Indigo carmine 610nm, and Bordeaux S 520nm). The concentration of urea was determined by absorbance 570nm, according to the urease-indophenol method (Urea NB, Wako Pure Chemical Industries, Ltd., Osaka, Japan). The concentration of glucose was determined by absorbance 505nm, according to the mutarotase-GOD method (Glucose C2, Wako Pure Chemical Industries, Ltd., Osaka, Japan).

6.2. Molecular-Size Screening

Remarkable size-screening capability was obtained between 60Da (Urea) and 604Da (Bordeaux S). The effective diffusion coefficient in the membrane D_{eff} decreased 2.5×10^4-fold even though the molecular-size increased only 10-fold [63] (Figure 11). This result strongly suggests that the mass-transfer channel was mono-disperse for molecular-sizes in our experiment. Wu and Imai reported that large dependence on molecular-size was achieved by specific polymer frameworks using pullulan and κ-carrageenan composite membranes [78].

The remarkable size-screening capability presented in Figure 11 was achieved by prepared 1.0M $CaCl_2$. The membrane composition was expressed as 0.1[mol-Ca^{2+}g-sodium alginate^{-1}], which is the molar ratio of molar Ca^{2+} to unit mass of alginate. The molar ratio of Ca^{2+} to alginate polymer is a dominant parameter of membrane preparation.

The diffusion coefficient in bulk aqueous phase D was plotted for comparison. It depended on the -0.6th power of the molecular weight. In contrast, the effective diffusion coefficient depended on almost the -5th power of the molecular weight of the tested components. The tested component did not adsorb to the membrane. The large dependence of the effective diffusion coefficient on molecular weight was due to the polymeric framework of a calcium alginate membrane, not due to adsorption. In contrast, the polymeric framework became dense to prepare the membrane (Figure 11).

6.3. Mass-Transfer Characteristics of Urea

The effective diffusion coefficient of urea (60Da) was evaluated mainly for mass-transfer characteristics as a typical small molecule.

6.3.1. Effect of Calcium Ion and F_{GG} on Mass-Transfer

The effective diffusion coefficient gradually decayed with increasing $CaCl_2$ concentration, due to the progress of cross-linking of molecular frameworks in the membrane (Figure 12). At $CaCl_2$ concentrations above 0.1M, the dependence of the effective diffusion coefficient on the $CaCl_2$ concentration became small [63]. This trend indicates that the molecular frameworks became saturated in this range, and that the effective diffusion coefficient remained almost constant. $CaCl_2$ acted as a cross-linker of molecular frameworks in the alginate molecular chain.

In the higher $CaCl_2$ concentration range, the effect of F_{GG} on the effective diffusion coefficient was especially remarkable (Figure 12). The polymeric structure of calcium alginate gels was governed mainly by intermolecular ionic bonds with homopolymeric blocks of the α-L-guluronic acid junction zone, in combination with Ca^{2+} [24].

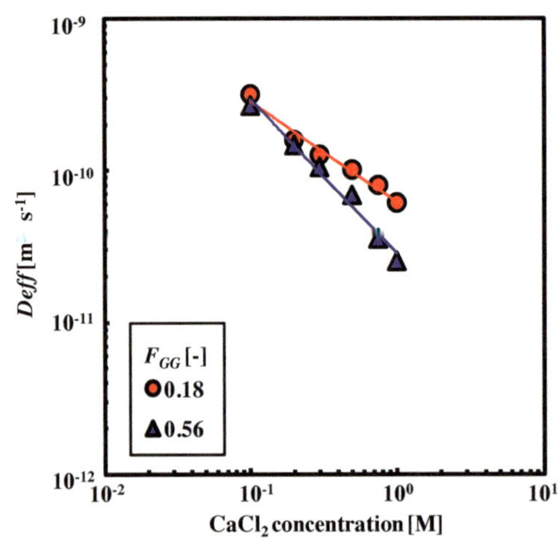

Figure 12. Effect of calcium chloride concentration on the effective diffusion coefficient of urea.

http://dx.doi.org/10.5772/50734

6.3.2. Effect of Water Content

The relationship between the effective diffusion coefficient and the volumetric water content of the membrane has been discussed using Eq. (8) by Yasuda's free volume theory [79].

$$\ln\left(\frac{D_{eff}}{D}\right) = -\frac{b(1-a)x}{1+ax} \tag{8}$$

Here, $x = (H_V^{-1}\text{-}1)$, $a = V_{fm}/V_{fl}$ and $b = V^*/V_{fl}$. H_V is the volumetric water fraction of the swollen membrane, a is the free volume ratio of the dry membrane (V_{fm}) to that of solvent (V_{fl}), and b is the volumetric ratio of the permeant characteristic volume (V *) to the free volume in the solvent (V_{fl}). D is the diffusion coefficient in bulk solvent calculated by the Wilke-Chang equation [80]. D_{eff} is the effective diffusion coefficient in the membrane. With this theory, ln (D_{eff} / D) is not generally a linear function of (H_V^{-1}-1).

There are two special cases of Eq. (8). First, ln (D_{eff} / D) becomes independent of membrane swelling at low H_V ($x\rightarrow\infty$). The left-hand term of Eq. (8) becomes almost constant. D_{eff} has a very low value. Second, for a region of high H_V ($x\rightarrow 0$), the effective diffusion coefficient is relatively large and decreases with decreasing H_V. In this case, ln (D_{eff} / D) is linearly proportional to (H_V^{-1}-1) and presents a negative slope of $b(1 - a)$.

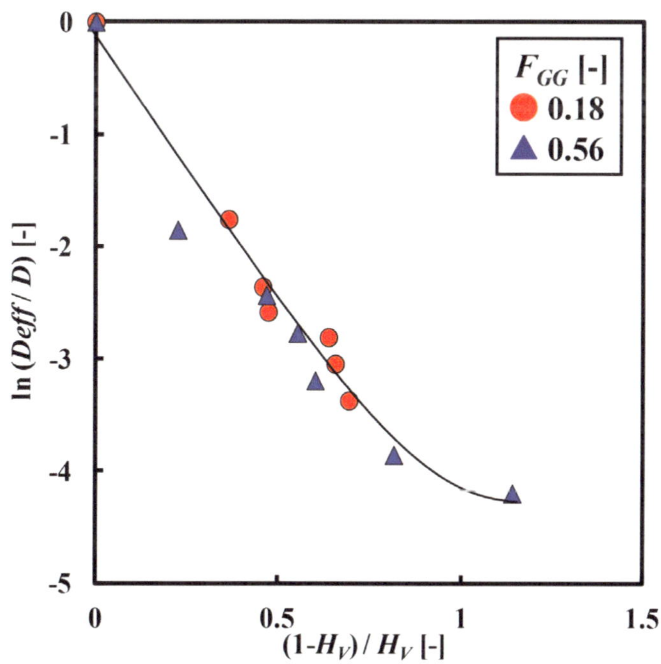

Figure 13. Effective diffusion coefficient of urea in calcium alginate membrane regulated by $CaCl_2$ concentration and F_{GG}, based on Yasuda's free volume theory.

Figure 13 depicts the ln (D_{eff} / D) of urea in a swollen calcium alginate membrane, based on the free volume theory (Eq. (8)). Here, ln (D_{eff} /D) vs. (H_V^{-1}-1) was linearly proportional with a negative slope. This trend has been reported for highly swollen membranes and/or very water-soluble solutes [81-82]. These two points were incorporated into our experiment conditions.

Ln (D_{eff} /D) vs. (H_V^{-1}-1) overlapped, regardless of having different F_{GG}. This result suggested that the value of (1 - a) and b are constant in our experiment range of F_{GG}.

(1 - a) represents the volumetric ratio of the void increased by membrane swelling due to the solvent. For urea transportation, the effect of F_{GG} on the free volume of the mass-transfer channel was not significant. In the future, the mass-transfer of other larger molecules in the membrane should be examined.

6.3.3. Tortuosity

The effective diffusion coefficient in porous materials can be represented by the following diffusion model. This model was applied to analyze mass-transfer in a swollen membrane.

$$D_{eff} = \frac{D\varepsilon}{\tau} \qquad (9)$$

Here ε is the void fraction and τ is the tortuosity of the membrane. The void fraction was assumed to be the volumetric water fraction of the swollen membrane (H_V) [74].

$$\tau = \frac{DH_V}{D_{eff}} \qquad (10)$$

The membrane tortuosity (τ) reflects the length of the mass-transfer channel compared to membrane thickness. Simple cylindrical mass-transfer channels across the membrane pass through at right angles to the membrane surface when tortuosity is unity (i.e., the average length of the channel is equivalent to membrane thickness). Channels usually take a more meandering path through the membrane; thus, typical tortuosities range from 1.5 to 2.5 [83].

Tortuosity of the calcium alginate membrane increased from 16 to 32 with increasing F_{GG}, which changed from 0.18 to 0.56 ($CaCl_2$ concentration of 1M). This result indicated that the mass fraction of α-L-guluronic acid was the dominant factor regulating tortuosity. However, the specific reason for a high level of tortuosity is not clear at present. Other factors inhibiting diffusion in the membrane could be speculated, (e.g., adsorption on the polymer network or molecular affinity between alginate polymer chains and the tested molecules) [65].

http://dx.doi.org/10.5772/50734

7. Water Permeation Flux

Water permeation flux is standard technical data for analyzing mass-transfer characteristics of the membrane [88]. Water permeation flux was evaluated based on the gravimetric or volumetric amount of water passing through the membrane. Gravimetric permeate flux (J_M) was generally calculated using the following equation.

$$J_M = \frac{M_P}{At} \tag{11}$$

Here, M_p is the permeate mass [kg], A is the membrane area [m^2], and t is the permeate time [s] [89]. Volumetric permeate flux (J_V) was calculated according to the following equation [90].

$$J_V = \frac{V_P}{At} \tag{12}$$

Here, V_P is the permeated volume of water (m^3), obtained from M_P / ρ_w. Wu and Imai investigated the water permeation flux of the pullulan-κ-carrageenan composite membrane [78].

7.1. Water Permeation Experiment

The typical procedure to measure water permeation flux is as follows. The permeation flux of the calcium alginate membrane was determined from the water mass flux using an ultrafiltration apparatus (UHP-62K, Advantec, Tokyo, Japan). The initial volume of feed solution was constant at 200ml. The operational pressure was adjusted by introducing nitrogen gas. The mass of permeated water was accurately measured by an electric balance and converted to volumetric water flux by recalculation using the density of the permeated water [63]. The experiment was carried out at 298K.

7.2. Effect of Cross-Linker Concentration on Water Permeation Flux

The water permeation flux decreased remarkably with increasing calcium chloride concentration due to progressive cross-linking of the molecular frameworks [63]. High water permeation flux was achieved with low $CaCl_2$ concentration as a cross-linker.

7.3. Dependence on Operational Pressure

Figure 14 illustrates the relationship between volumetric water flux and operational pressure ΔP on the calcium alginate membrane prepared by 1M $CaCl_2$. The water permeation flux increased almost linearly with increasing operational pressure. The water permeation mechanism was assumed to be Hagen-Poiseuille flow. High water permeation flux was realized in the low F_{GG} (0.18) membrane.

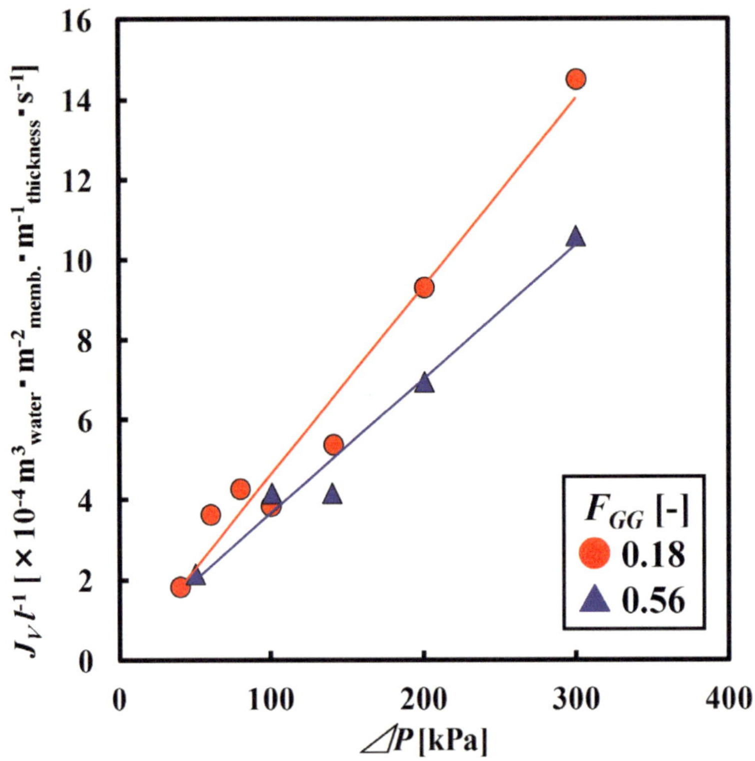

Figure 14. Water permeation flux of calcium alginate membrane ($CaCl_2$: 1M) prepared from different F_{GG} by applying different pressures.

7.4. Water Permeation Flux of Other Membranes

Table 4 presents previous investigation results of the water permeation flux of various membranes. The pure water flux of the calcium alginate membrane at ⊿ P 20 [kPa] was obtained as 9.3 $\times 10^{-9}$ [m^3 m^{-2} s^{-1}], which is lower than that of chitosan [45], cellulose acetate [91], and cellulose acetate with polyethylene glycol (PEG) [91]. It was assumed that the polymer framework of the calcium alginate membrane became remarkably dense, which led to decreasing water permeation flux.

Ethanol aqueous solution was previously examined in pervaporation using a sodium alginate membrane cross-linked by phosphoric acid. The permeation flux (1.3×10^{-5} kg m^{-2} s^{-1}) was less than that of the PTMSP (poly (1-trimethylsilyl-1-propyne)) membrane (5.8×10^{-5} kg m^{-2} s^{-1}) [55, 92]. The sodium alginate membrane blended with dextrin was cross-linked with glutaraldehyde to make a stable membrane. It exhibited better water permeation flux of isopropanol aqueous solution than the PVA coating alginate membrane [58, 93]. The water permeation flux was improved with the use of the PVA single component membrane presented by Naidu et al. [94].

http://dx.doi.org/10.5772/50734

Authers	Membrane	Driving mechanism	Permeated solution	Water concentration	P [kPa]	J_M [kg m^{-2} s^{-1}]	J_V [m^3 m^{-2} s^{-1}]	Rf.
Kashima et al.	Calcium alginate	Pressure	Pure water	-	20	-	9.3×10^{-9}	63
Takahashi et al.	Chitosan	Pressure	Pure water	-	20	-	3.4×10^{-8}	45
Saljoughi et al.	Cellulose acetate	Pressure	Pure water	-	20	-	2.2×10^{-6}	91
	Cellulose acetate - PEG		Pure water	-	20	-	6.1×10^{-5}	
Kalyani et al.	Sodium alginate	Pervaporation	Ethanol + Water	9.74 wt.%	6.7×10^{-3}	1.3×10^{-5}	-	55
Claes et al.	PTMSP	Pervaporation	Ethanol + Water	10 wt.%	4×10^{-3}	5.8×10^{-5}	-	92
	PTMSP - silica		Ethanol + Water	10 wt.%	4×10^{-3}	1.6×10^{-4}	-	
Saraswahi et al.	Sodium alginate	Pervaporation	Isopropanol + Water	10 wt.%	1.33	1.9×10^{-5}	-	58
	Sodium alginate-Dextrin		Isopropanol + Water	10 wt.%	1.33	3.7×10^{-5}	-	
Kurkuri et al.	Sodium alginate - PVA (75:25)	Pervaporation	Isopropanol + Water	10 wt.%	1.33	6.7×10^{-6}	-	93
	Sodium alginate - PVA (50:50)		Isopropanol + Water	10 wt.%	1.33	9.4×10^{-6}	-	
	Sodium alginate - PVA (25:75)		Isopropanol + Water	10 wt.%	1.33	1.1×10^{-5}	-	
Naidu et al.	PVA	Pervaporation	Isopropanol + Water	10 wt.%	1.33	2.6×10^{-5}	-	94
	PVA - Polyaniline		Isopropanol + Water	10 wt.%	1.33	1.9×10^{-5}	-	

PEG: polyethylene glycol, PTMSP: Poly (1-trimethylsilyl-1-propyne), PVA: Polyvinyl alcohol

Table 4. Water permeation flux of pressure-derived permeation (J_V) and pervaporation (J_M).

8. Conclusion

Advanced membrane material from marine biological polymer and sensitive molecular-size recognition for promising separation technology were demonstrated. Stable calcium alginate membrane in swollen state was successfully prepared. The calcium alginate membrane has better mechanical properties than other biopolymer membranes for conventional use. The calcium alginate membrane has a high void fraction (50% to 90%) similar to commercial microfiltration membrane (60% to 90%). Mass transfer characteristics are evidently changed by the mass fraction of α-L-guluronic acid (F_{GG}) and additive $CaCl_2$. Water permeation flux of the calcium alginate membrane is lower than that of other biopolymer membrane (e.g. chitosan, cellulose). In future, the water permeation flux is improved by combination with other polymers (e.g. dextrin). Alginate membrane should be developed as an alternative to artificial polymer membranes.

Acknowledgements

Sodium alginate samples I-2M and I-2G were kindly provided by KIMICA Corporation (Tokyo, Japan). Professor Hirosi Anzai of Nihon University kindly provided much useful advice for determining the F_{GG} in sodium alginate. Dr. Naoto Arai of Nihon University kindly provided technical advice for SPM observation. Dr. Kei Tao of Nihon University kindly provided technical support of instruments. The authors are grateful to them.

Author details

Keita Kashima and Masanao Imai*

*Address all correspondence to: XLT05104@nifty.com

Course in Bioresource Utilization Sciences, Graduate School of Bioresource Sciences, Nihon University, Japan

References

[1] Sokolnicki, A. M., Fisher, R. J., Harrah, T. P., & Kaplan, D. L. (2006). Permeability of bacterial cellulose membranes. *J. of Membrane Science*, 272, 15-27.

[2] Zhao, M., Xu, X-L., Jiang, Y-D., Sun, W-Z., Wang, W-F., & Yuan, L-M. (2009). Enantioseparation of trans-stilbene oxide using a cellulose acetate membrane. *J. of Membrane Science*, 336, 149-153.

[3] Bigi, A., Cojazzi, G., Panazavolta, S., Roveri, N., & Rubini, K. (2002). Stabilization of gelatin films by crosslinking with genipin. *Biomaterials*, 23, 4827-4832.

[4] Takahashi, T., Imai, M., & Suzuki, I. (2007). Water permeability of chitosan membrane involved in deacetylation degree control. *Biochemical Engineering J.*, 36, 43-48.

[5] Vörösmarty, C. J., Green, P., Salisbury, J., & Lammers, R. B. (2000). Global Water Resources: Vulnerability from Climate Change and Population Growth. *Science*, 289, 284-288.

[6] Gilau, A. M., & Small, M. J. (2008). Designing Cost-Effective Sea Water Reverse Osmosis System under Optimal Energy. *Renewable Energy*, 33, 617-630.

[7] Charcosset, C. (2009). A review of membrane processes and renewable energies for desalination. *Desalination*, 245, 214-231.

[8] Hsu, S. T., Cheng, K. T., & Chiou, J. S. (2002). Seawater desalination by direct contact membrane distillation. *Desalination*, 143, 279-287.

[9] Peng, P., Fane, A. G., & Li, X. (2005). Desalination by membrane distillation adopting a hydrophilic membrane. *Desalination*, 173, 45-54.

[10] Gazagnes, L., Cerneaux, S., Persin, M., Prouzet, E., & Larbot, A. (2007). Desalination of sodium chloride solutions and seawater with hydrophobic ceramic membranes. *Desalination*, 217, 260-266.

[11] Ye, Y., Clech, P. L., Chen, V., Fane, A. G., & Jefferson, B. (2005). Fouling mechanisms of alginate solutions as model extracellular polymeric substances. *Desalination*, 175, 7-20.

[12] Katsoufidou, K., Yiantsios, S. G., & Karabelas, A. J. (2008). An experimental study of UF membrane fouling by humic acid and sodium alginate solutions: the effect of backwashing on flux recovery. *Desalination*, 220, 214-227.

[13] Kashima, K., Imai, M., & Suzuki, I. (2009). Molecular sieving character of calcium alginate membrane relate with polymeric framework and mechanical strength. *J. Bioscience and Bioengineering*, 108, 69.

[14] Papageorgiou, S. K., Katsaros, F. K., Favvas, E. P., Romanos, G. E., Athanasekou, C. P., Beltsios, K. G., Tzialla, O. I., & Falaras, P. (2012). Alginate fibers as photocatalyst immobilizing agents applied in hybrid photocatalytic/ultrafiltration water treatment processes. *Water Research*, 46, 1858-1872.

[15] Takahashi, T., Imai, M., Suzuki, I., & Sawai, J. (2008). Growth inhibitory effect on bacteria of chitosan membranes regulated with deacetylation degree. *Biochemical Engineering J.*, 40, 485-491.

[16] Miao, J., Chen, G., Gao, C., & Dong, S. (2008). Preparation and characterization of N,O-carboxymethyl chitosan/Polysulfone composite nanofiltration membrane crosslinked with epichlorohydrin. *Desalination*, 233, 147-156.

[17] Padaki, M., Isloor, A. M., Fernandes, J., & Prabhu, K. N. (2011). New polypropylene supported chitosan NF-membrane for desalination application. *Desalination*, 280, 419-423.

[18] Haddad, R., Ferjani, E., Roudesli, M. S., & Deratani, A. (2004). Properties of cellulose acetate nanofiltration membranes. Application to brackish water desalination. *Desalination*, 167, 403-409.

[19] Zhang, S., Wang, K. Y., Chung, T-S., Jean, Y. C., & Chen, H. (2011). Molecular design of the cellulose ester-based forward osmosis membranes for desalination. *Chemical Engineering Science*, 66, 2008-2018.

[20] Semiat, R. (2008). Energy issues in desalination processes. *Environmental Science & Technology*, 42, 8193-8201.

[21] Stanford, E. C. C. (1881). British Patent: 142.

[22] Cottrell, I. W., & Kovacs, P. (1980). Alginate. Davidson R T, editor, *Handbook of Water-Soluble Gums and Resins*, United States of America, McGraw-Hill, Inc., 2-1-2-43.

[23] Julian, T. N., Radebaugh, G. W., & Wisniewski, S. J. (1988). Permeability characteristics of calcium alginate films. *J. Controlled Release*, 7, 165-169.

[24] Gacesa, P. Alginates. 1988, *Carbohydrate Polymers*, 8, 161-182.

[25] Lee, S., Ang, W. S., & Elimelech, M. (2006). Fouling of reverse osmosis membranes by hydrophilic organic matter: implications for water reuse. *Desalination*, 187, 313-321.

[26] Ang, W. S., Lee, S., & Elimelech, M. (2006). Chemical and physical aspects of cleaning of organic-fouled reverse osmosis membranes. *J. Membrane Science*, 272, 198-210.

[27] Lee, S., & Elimelech, M. (2006). Relating organic fouling of reverse osmosis membranes to intermolecular adhesion forces. *Environmental Science & Technology*, 40(3), 980-987.

[28] Haug, A., Larsen, B., & Smidsrød, O. (1974). Uronic acid sequence in alginate from different sourcese. *Carbohydrate Research*, 32, 217-225.

[29] Atkins, E. D. T., Nieduszynski, I. A., Mackie, W., Parker, K. D., & Smolko, E. E. (1973). Structural components of alginic acid. I. The crystalline structure of Poly-β-D-mannuronic acid. Results of X-Ray diffraction and polarized infrared studies. *Biopolymers*, 12, 1865-1878.

[30] Atkins, E. D. T., Nieduszynski, I. A., Mackie, W., Parker, K. D., & Smolko, E. E. (1973). Structural components of alginic acid. II. The crystalline structure of Poly-α-L-guluronic. Results of X-Ray diffraction and polarized infrared studies. *Biopolymers*, 12, 1879-1887.

[31] Campa, C., Holtan, S., Nilsen, N., Bjerkan, T. M., Stokke, B. T., & Skjåk-bræk, G. (2004). Biochemical analysis of the processive mechanism for epimerization of alginate by mannuronan C-5 epimerase AlgE4. *Biochemical J.*, 381, 155-164.

[32] Donati, I., Holtan, S., Mørch, Y. A., Borgogna, M., Dentini, M., & Skjåk-bræk, G. (2005). New Hypothesis on the Role of Alternating Sequences in Calcium-Alginate Gels. *Biomacromolecules*, 6, 1031-1040.

[33] Strand, B. L., Mørch, Y. A., Syvertsen, K. R., Espevik, T., & Skjåk-bræk, G. (2003). Microcapsules made by enzymatically tailored alginate. *J. Biomedical Materials Research*, 64A, 540-550.

[34] Katsoufidou, K., Yiantsios, S. G., & Karabelas, A. J. (2007). Experimental study of ultrafiltration membrane fouling by sodium alginate and flux recovery by backwashing. *J. Membrane Science*, 300, 137-146.

[35] Konsoula, Z., & Liakopoulou-Kyriakides, M. (2006). Starch hydrolysis by the action of an entrapped in alginate capsules α-amylase from Bacillus subtilis. *Process Biochemistry*, 41, 343-349.

[36] Almeida, P. F., & Almeida, A. J. (2004). Cross-linked alginate-gelatine beads: a new matrix for controlled release of pindolol. *J. of Controlled Release*, 97, 431-439.

[37] Becerra, M., Baroli, B., Fadda, A. M., Blanco Me´ndez, J., & Gonza´lez Siso, M. J. (2001). Lactose bioconversion by calcium-alginate immobilization of Kluyveromyces lactis cells. *Enzyme and Microbial Technology*, 29, 506-512.

[38] Milovanovic, A., Bozic, N., & Vujcic, Z. (2007). Cell wall invertase immobilization within calcium alginate beads. *Food Chemistry*, 104, 81-86.

[39] Dembczynski, R., & Jankowski, T. (2000). Characterisation of small molecules diffusion in hydrogel-membrane liquid-core capsules. *Biochemical Engineering J.*, 6, 41-44.

[40] Smidsrød, O., & Haug, A. (1968). Dependence upon Uronic Acid Composition of Some Ion-Exchange Properties of Alginates. *Acta Chemica Scandinavica*, 22, 1989-1997.

[41] Draget, K. I., Smidsrød, O., & Skjåk-Bræk, G. (2000). Alginates from Algae. In: Stenbuchel A, Rhee S K, editors, *Polysaccharides and Polyamides in the Food Industry. Properties, Production and Patents*, Federal Republic of Germany, Wiley-VCH, 1-30.

[42] Grant, G. T., Morris, E. R., Rees, D. A., Smith, P. J. C., & Thom, D. (1973). Biological interactions between polysaccharides and divalent cations: The egg-box model. *FEBS Letters*, 32, 195-198.

[43] Kvam, B. J., Grasdalen, H., Smidsrød, O., & Anthonsen, T. (1986). NMR Studies of the Interaction of Metal Ions with Poly(1,4-hexuronates). VI. Lanthanide(III) Complexes of Sodium (Methyl alpha-D-galactopyranosid) uronate and Sodium (Phenylmethyl alpha-D-galactopyranosid) uronate. *Acta Chemica Scandinavica*, B40, 735-739.

[44] Sossna, M., Hollas, M., Schaper, J., & Scheper, T. (2007). Structural development of asymmetric cellulose acetate microfiltration membranes prepared by a single-layer dry-casting method. *J. of Membrane Science*, 289, 7-14.

[45] Takahashi, T., Imai, M., & Suzuki, I. (2008). Cellular structure in an N-acetyl-chitosan membrane regulate water permeability. *Biochemical Engineering J.*, 42, 20-27.

[46] Teixeira, J. A., Mota, M., & Venancio, A. (1994). Model identification and diffusion coefficients determination of glucose and malic acid in calcium alginate membranes. *The Chemical Engineering J.*, 56, B9-B14.

[47] Zhang, W., & Franco, C. M. M. (1999). Critical assessment of quasi-steady-state method to determine effective diffusivities in alginate gel membranes. *Biochemical Engineering J.*, 4, 55-63.

[48] Grassi, M., Colombo, I., & Lapasin, R. (2001). Experimental determination of the theophylline diffusion coefficient in swollen sodium-alginate membranes. *J. Controlled Release*, 76, 93-105.

[49] Hubble, J., & Newman, J. D. (1985). Alginate Ultrafiltration Membranes. *Biotechnology Letters*, 7(4), 273-276.

[50] Andreopoulos, A. G. (1987). Diffusion characteristics of alginate membranes. *Biomaterials*, 8, 397-400.

[51] Aslani, P., & Kennedy, R. A. (1996). Studies on diffusion in alginate gels. I. Effect of cross-linking with calcium or zinc ions on diffusion of acetaminophen. *J. Controlled Release*, 42, 75-82.

[52] Zimmermann, H., Wählisch, F., Baier, C., Westhoff, M., Reuss, R., Zimmermann, D., Behringer, M., Ehrhart, F., Katsen-Globa, A., Giese, C., Marx, U., Sukhorukov, V. L., Vásquez, J. A., Jakob, P., Shirley, S. G., & Zimmermann, U. (2007). Physical and biological properties of barium cross-linked alginate membranes. *Biomaterials*, 28, 1327-1345.

[53] Toti, U. S., & Aminabhavi, T. M. (2004). Different viscosity grade sodium alginate and modified sodium alginate membranes in pervaporation separation of water + acetic acid and water + isopropanol mixtures. *J. of Membrane Science*, 228, 199-208.

[54] Yeom, C. K., & Lee, K-H. (1998). Characterization of Sodium Alginate Membrane Crosslinked with Glutaraldehyde in Pervaporation Separation. *J. of Applied Polymer Science*, 67, 209-219.

[55] Kalyani, S., Smitha, B., Sridhar, S., & Krishnaiah, A. (2008). Pervaporation separation of ethanol-water mixtures through sodium alginate membranes. *Desalination*, 229, 68-81.

[56] Yang, G., Zhang, L., Peng, T., & Zhong, W. (2000). Effects of Ca^{2+} bridge cross-linking on structure and pervaporation of cellulose/alginate blend membranes. *J. of Membrane Science*, 175, 53-60.

[57] Chen, J. H., Ni, J. C., Liu, Q. L., & Li, S. X. (2012). Adsorption behavior of Cd(II) ions on humic acid-immobilized sodium alginate and hydroxyl ethyl cellulose blending porous composite membrane adsorbent. *Desalination*, 285, 54-61.

[58] Saraswathi, M., Rao, K. M., Prabhakar, M. N., Prasad, C. V., Sudakar, K., Naveen Kumar, H. M. P., Prasad, M., Rao, K. C., & Subha, M. C. S. (2011). Pervaporation studies of sodium alginate (SA)/dextrin blend membranes for separation of water and isopropanol mixture. *Desalination*, 269, 177-183.

[59] Wang, X. P. (2000). Modified alginate composite membranes for the dehydration of acetic acid. *J. of Membrane Science*, 170, 71-79.

[60] Kanti, P., Srigowri, K., Madhuri, J., Smitha, B., & Sridhar, S. (2004). Dehydration of ethanol through blend membranes of chitosan and sodium alginate by pervaporation. *Separation and Purification Technology*, 40, 259-266.

[61] Reddy, A. S., Kalyani, S., Kumar, N. S., Boddu, V. M., & Krishnaiah, A. (2008). Dehydration of 1,4-dioxane by pervaporation using crosslinked calcium alginate-chitosan blend membranes. *Polymer Bulletin*, 61, 779-790.

[62] Smitha, B., Sridhar, S., & Khan, A. A. (2005). Chitosan-sodium alginate polyion complexes as fuel cell membranes. *European Polymer J.*, 41, 1859-1866.

[63] Kashima, K., Imai, M., & Suzuki, I. (2010). Superior molecular size screening and mass-transfer characterization of calcium alginate membrane. *Desalination and Water Treatment*, 17, 143-149.

[64] Li, Y., Hu, M., Du, Y., Xao, H., & McClements, D. J. (2011). Control of lipase digestibility of emulsified lipids by encapsulation within calcium alginate beads. *Food Hydrocolloids*, 25, 122-130.

[65] Kashima, K., & Imai, M. (2011). Dominant impact of the α-L-guluronic acid chain on regulation of the mass transfer character of calcium alginate membranes. *Desalination and Water Treatment*, 34, 257-265.

[66] Bitter, T., & Muir, H. M. (1962). A modified uronic acid carbazole reaction. *Analytical Biochemistry*, 4, 330-334.

[67] Anzai, H., Uchida, N., & Nishide, E. (1986). Comparative studies of colorimetric analysis for uronic acids. *Bull. Coll. Agr. & Vet. Med., Nihon Univ.*, 43, 53-56.

[68] Benavides, S., Villalobos-Carvajal, R., & Reyes, J. E. (2012). Physical, mechanical and antibacterial properties of alginate film: Effect of the crosslinking degree and oregano essential oil concentration. *J. Food Engineering*, 110, 232-239.

[69] Rhim, J-W. (2004). Physical and mechanical properties of water resistant sodium alginate films. *Lebensmittel-Wissenschaft und-Technologie*, 37, 323-330.

[70] Olivas, G. I., & Barbosa-Canovas, G. V. Alginate-calcium films: Water vapor permeability and mechanical properties as affected by plasticizer and relative humidity. 2008, *LWT- Food Science and Technology*, 41, 359-366.

[71] Boricha, A. G., & Murthy, Z. V. P. (2010). Preparation of N,O-carboxymethyl chitosan/cellulose acetate blend nanofiltration membrane and testing its performance in treating industrial wastewater. *Chemical Engineering Journal*, 157, 393-400.

[72] Huang, Q-L., Xiao, C-F., Hu, X-Y., & Li, X-F. (2011). Study on the effects and properties of hydrophobic poly(tetrafluoroethylene) membrane. *Desalination*, 277, 187-192.

[73] Schwarz, H-H., & Hicke, H-G. (1989). Influence of Casting Solution Concentration on Structure and Performance of Cellulose Acetate Membranes. *J. Membrane Science*, 46, 325-334.

[74] So, M. T., Eirich, F. R., Strathmann, H., & Baker, R. W. (1973). Preparation of Asymmetric Loeb-sourirajan Membranes. *J. Polymer Science: Polymer Letters Edition*, 11, 201-205.

[75] Block, W. (2003). Water status and thermal analysis of alginate beads used in cryopreservation of plant germplasm. *Cryobiology*, 47, 59-72.

[76] Al-Rub, F. A. A., Banat, F., & Beni-Melhim, K. (2002). Parametric sensitivity analysis of direct contact membrane distillation. *Separation Science and Technology*, 37(14), 3245-3271.

[77] Adnan, S., Hoang, M., Wang, H., & Xie, Z. (2012). Commercial PTFE membranes for membrane distillation application: Effect of microstructure and support material. *Desalination*, 284, 297-308.

[78] Wu, P., & Imai, M. (2011). Food polymer pullulan-κ-carrageenan composite membrane performed smart function both on mass transfer and molecular size recognition. *Desalination and Water Treatment*, 34, 239-245.

[79] Yasuda, H., Lamaze, C. E., & Peterlin, A. (1971). Diffusive and hydraulic permeabilities of water in water-swollen polymer membranes. *J. Polymer Science: Part A-2*, 9, 1117-1131.

[80] Wilke, C. R., & Chang, P. (1955). Correlation of Diffusion Coefficients in Dilute Solutions. *AIChE J.*, 1, 264-270.

[81] Chen, S. X., & Lostritto, R. T. (1996). Diffusion of benzocaine in poly (ethylene-vinyl acetate) membranes: Effects of vehicle ethanol concentration and membrane vinyl acetate content. *J. Controlled Release*, 38, 185-191.

[82] Vadalkar, V. S., Kulkarni, M. G., & Bhagwat, S. S. (1993). Anomalous sorption of binary solvents in glassy polymers: interpretation of solute release at constant rates. *Polymer*, 34, 4300-4306.

[83] Baker, R. W. (2004). *Membrane Technology and Applications*, England, John Wiley & Sons, Ltd., 67-68.

[84] Mehmetoglu, U. (1990). Effective diffusion coefficient of sucrose in calcium alginate gel. *Enzyme Microbial Technology*, 12, 124-126.

[85] Longo, M. A., Novella, I. S., Garcia, L. A., & Diaz, M. (1992). Diffusion of proteases in calcium alginate beads. *Enzyme and Microbial Technology*, 14, 586-590.

[86] Kikuchi, A., Kawabuchi, M., Watanabe, A., Sugihara, M., Sakurai, Y., & Okano, T. (1999). Effect of Ca^{2+}-alginate gel dissolution on release of dextran with different molecular weights. *J. Controlled Release*, 58, 21-28.

[87] Koyama, K., & Seki, M. (2004). Evaluation of Mass-Transfer Characteristics in Alginate-Membrane Liquid-Core Capsules Prepared Using Polyethylene Glycol. *J. Bioscience and Bioengineering*, 98, 114-121.

[88] Merdaw, A. A., Sharif, A. O., & Derwish, G. A. W. (2010). Water permeability in polymeric membranes, Part I. *Desalination*, 260, 180-192.

[89] Cho, C. H., Oh, K. Y., Kim, S. K., Yeo, J. G., & Sharma, P. (2011). Pervaporative seawater desalination using NaA zeolite membrane: Mechanisms of high water flux and high salt rejection. *J. Membrane Science*, 371, 226-238.

[90] Hsieh, K. H., Lin, B. Y., & Chiu, W. Y. (1989). Studies on Diisocyanate-Modified Cellulose Acetate Membranes. *Desalination*, 71, 97-105.

[91] Saljoughi, E., Sadrzadeh, M., & Mohammadi, T. (2009). Effect of preparation variables on morphology and pure water permeation flux through asymmetric cellulose acetate membranes. *J Membrane Science*, 326, 627-634.

[92] Claes, S., Vandezande, P., Mullens, S., Leysen, R., De Sitter, K., Andersson, A., Maurer, F. H. J., Van den Rul, H., Peeters, R., & Van Bael, M. K. (2010). Pervaporation separation of water + isopropanol mixtures using novel nanocomposite membranes of poly(vinyl alcohol) and polyaniline. *J Membrane Science*, 351, 160-167.

[93] Kurkuri, M. D., Toti, U. S., & Aminabhavi, T. M. (2002). Syntheses and Characterization of Blend Membranes of Sodium Alginate and Poly(vinyl alcohol) for the Pervaporation Separation of Water + Isopropanol Mixtures. *J Applied Polymer Science*, 86, 3642-3651.

[94] Vijaya Kumar Naidu, B., Sairam, M., Raju, K. V. S. N., & Aminabhavi, T. M. (2005). High flux composite PTMSP-silica nanohybrid membranes for the pervaporation of ethanol/water mixtures. *J. Membrane Science*, 260, 142-155.

Chapter 2

Desalination of Industrial Effluents Using Integrated Membrane Processes

Marek Gryta

Additional information is available at the end of the chapter

http://dx.doi.org/10.5772/49993

1. Introduction

The term integrated or hybrid membrane processes refers to the integration of one or more membrane processes with or without the conventional unit operations in order to increase the performance depending on the type of feed and product quality required [1, 2]. A coupling of different membrane processes into the integrated systems allowed to develop the efficient technologies, which have been successfully used for water desalination and the treatment of industrial effluents [1-3]. The realization of these processes resulted in the production of fresh or process water. The most current techniques of desalination are thermal distillation and the membranes technology: electrodialysis (ED) and reverse osmosis (RO) [2].

The installation for the desalination of water and different effluents create a certain amount of the concentrates that must be disposed in an environmentally appropriate manner. Each year, a large amount of RO concentrates is discharged and leads to a significant loss of water resource and disposal challenge. In spite of the scale of this economical and environmental problem, the options for brine management for the inland plants are rather limited. These options include: discharge to surface water or wastewater treatment plants, deep well injection, land disposal, evaporation ponds, and mechanical/thermal evaporation [1-3].

The diluted solutions of salts are expensive for the treatment; therefore, in the case of industrial effluents they are often discharged directly into the environment. The discharges of such wastewater pose an environmental impact and cause an enhancement in the surface water salinity [2, 4]. The separation of salt from aqueous solutions is carried out most frequently using the thermal methods, such as multi-stage flash evaporation (MSF). The thermal processes are based on improved distillation, evaporation and condensation technologies with the aim to save energy and to obtain fresh water with a low level of TDS and at lower operating costs. However, with regard to energy consumption, these processes are essentially used for the concentration of brines.

The thermal processes are generally more expensive than RO, although a distillation method produces pure water independently on the quality and salinity of the feed water. Reverse osmosis can be employed for a preliminary concentration of saline wastewater, thus the cost of the thermal methods will be considerably reduced and the range of their application can include the diluted solutions. However, the RO process requires the application of a sophisticated pre-treatment due to the considerable problems associated with the fouling and scaling [2]. The applications of nanofiltration (NF) or ultrafiltration/nanofiltration (UF/NF) integrated system, to achieve the removal of the divalent ions, allows to limit the scaling phenomenon (mainly $CaSO_4$) and the concentration factor in the RO process can be enhanced [2-5].

The effluents containing salts together with several organic substances, such as protein or polysaccharides, are generated during the realization of several industrial processes [2-6]. The biological methods are traditionally used for wastewater treatment. A high concentration of salt in wastewater possesses additional problems in its purification. Moreover, current and pending state regulations lower effluent discharge limits, therefore these methods cannot meet new restrictions. The best solution is realization the zero liquid discharge idea, which means that the industrial effluents are recovered as a clean stream for re-use, and the concentrate stream can be disposed off in an environmentally safety manner, or further reduced to solid [2, 4, 6, 7]. The RO concentrates can be treated with ED, which is capable of achieving a maximum concentrate of 80 g/L, and evaporators achieve a concentration of 300 g/L [2].

The solutions containing the large amounts of salts can be concentrated using the membrane distillation (MD) [6-8]. This process has a potential application for the water desalination and the wastewater treatment [6, 9, 10]. The separation mechanism of the MD process is based on the vapour/liquid equilibrium of liquid mixtures. The volatile components of the feed evaporate through the pores of the membrane; therefore, the presence of the vapour phase in the pores is a necessary condition for MD [11, 12]. During the MD process of the solutions containing non-volatile solutes, only the water vapour is transferred through the non-wetted hydrophobic porous membrane, and the obtained distillate comprises high purity water. The MD process enables the production of pure water from water solutions, the quality of which impedes a direct application of the RO for this purpose [6, 7, 11-13].

In the direct contact MD variant (DCMD) the membrane separates the hot feed from the cold distillate (Figure 1). The driving force for the mass transport is a difference in the vapour pressures, resulting from different temperatures and the compositions of the solutions in the layers adjacent to the membrane [11, 14].

http://dx.doi.org/10.5772/49993

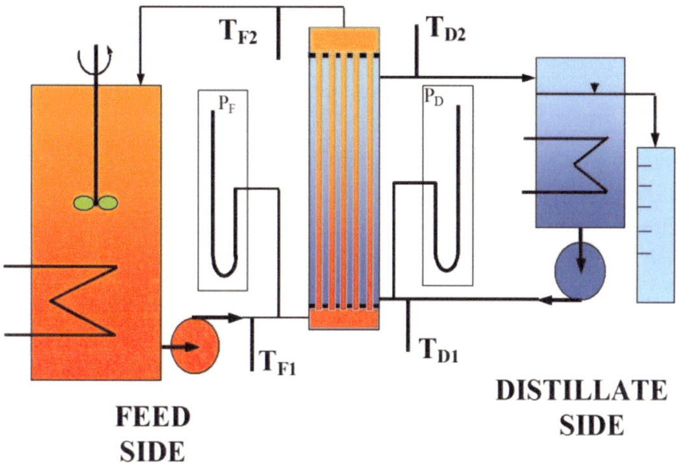

Figure 1. Experimental set-up of DCMD installation.

The treatment of saline wastewater can be performed utilizing the MD process to produce pure water and a concentrate containing the substances present in the parent solution. Subsequently, the components of this concentrate can be separated using the crystallisation of salt followed by the precipitation of dissolved substances [7, 8, 14-16]. This enables the disposal of such complex wastewater or their recycling. The obtained process water can be recycled for technological purposes.

The industrial use of the vacuum membrane distillation (VMD) to further concentrate the RO brines was proposed as a complementary process to seawater desalination [10]. Although, scaling occurs in VMD for high salt concentrations, its impact on the permeate flux was very limited. Large areas of the membranes remain free of visible fouling. The main salts responsible for the scaling are calcium crystals such as calcium carbonate and calcium sulphate, which have the lowest solubility. In order to reduce the scaling, the accelerated precipitation softening was integrated with direct contact membrane distillation, which establish a desalination process for high-recovery desalting of primary reverse osmosis concentrate [7, 17].

The integrated membrane processes were implemented in the industry for water re-use, therefore, the volume of discharged wastewater was significantly reduced [1, 2]. Moreover, the membrane technologies are presently an established part of several industrial processes. The use of membranes to separate and recover products and by-products from process streams may increase the process efficiency [2, 5, 18, 19]. The possibility of application of integrated membrane systems, especially the MD process combined with crystallization, created one of the new possibilities to solve the saline wastewater treatment [14-17].

2. Effluents concentration integrated with salt precipitation

A problem associated with effluents management may results from two reasons: a) the generated solutions are of appropriate purity, but the concentration of solute is too low to find a practical application; b) a solution contains, besides desired components, also impurities which preclude its further utilization. In both cases, the concentration or treatment of effluents by traditional methods is generally unprofitable [2, 15]. The concentration of such solutions by membrane processes, often in combination with the salt crystallization can be an advantageous alternative [4, 7, 8].

The membrane technologies were used for processing a wide spectrum of feed streams containing salts in wide variety of applications. The implementation of integrated membrane systems appears an interesting possibility for improving desalination operations [2]. For example, the RO, NF and DCMD processes were used for the removal of boron and arsenic from water [20]. A serious problem is associated with the presence besides the ions such salts like NaCl, $FeCl_3$ or $CuSO_4$ also the Ca^{2+} and HCO_3^- ions in the treated solutions. As a result, the precipitation of $CaSO_4$ and $CaCO_3$ can occurs during the separation of effluents [21]. It is referred as „scaling", and takes place in both membranes processes (Figure 2) and the heat exchangers (evaporators) installations [22].

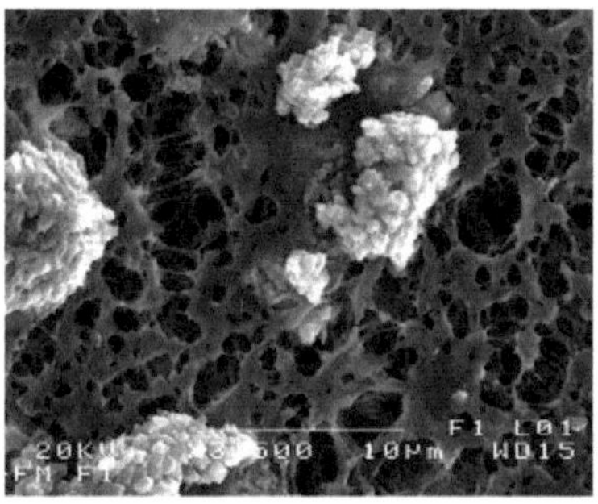

Figure 2. SEM image of $CaCO_3$ deposit precipitated during MD process. Feed: tap water.

It was demonstrated that the induction times are significantly reduced if the precipitation take place in the presence of membranes, thus causing an increase of the $CaCO_3$ nucleation rate [7]. However, the MD process is affected by bicarbonates dissolved in water, which undergo a decomposition reaction when feed is heated and the precipitation of carbonates on the membranes surface was observed [23-25]. As a consequence, during the first few hours of water desalination the $CaCO_3$ deposit layer was formed on the membrane surface (Figure 3A). The deposit was significantly reduced by means of either feeding MD installations with water pretreated in an accelerator (softening and coagulation) [26] or by acidifying them up to pH=4 [13]. Although the chemical coagulation followed by sand filtration or microfiltration decreases the fouling potential, a deposition of small amounts of precipitate on the membrane surface is still observed (Figure 3B).

In the case of the membranes, the deposit is formed not only onto the surface, but also penetrates their pores interior, what may damage the membrane structure [21]. Such properties exhibit especially the crystals of gypsum ($CaSO_4$ $2H_2O$) (Figure 4). The crystallites inside the membrane occupy the space significantly larger than the pore dimensions, what leads to a destruction of the membrane structure. The application of membranes that are significantly thicker than a layer of crystallizing salt allows to maintain the mechanical strength of the membranes and prevents the membranes from complete wetting during MD process.

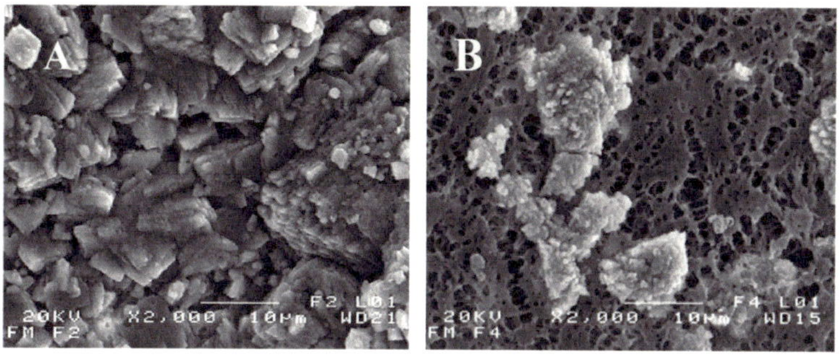

Figure 3. SEM images of deposit formed on the membrane surface during the water desalination by MD process. A) desalination of natural water, B) desalination of industrially pre-treated water (accelator).

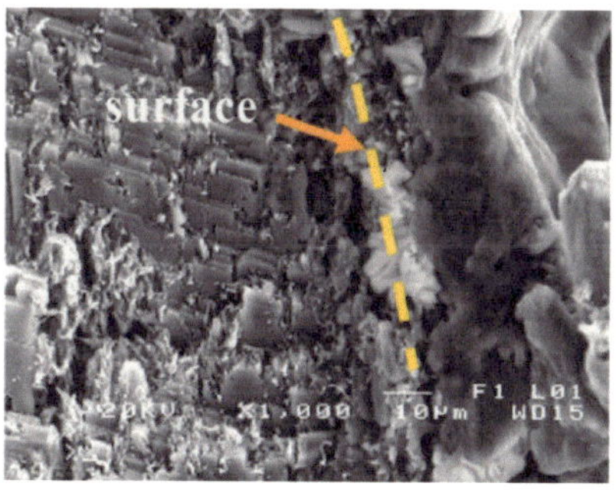

Figure 4. SEM image of membrane cross-section with $CaSO_4$ deposit formed on the membrane surface during MD process of brackish water.

The formation of deposit during effluents desalination can be limited by adding antiscalant to the feed solutions. The application of polyphosphates as antiscalant restricted the amount of deposits formed on the membrane surface during the desalination of water by MD process. However, the morphology of deposit was changed, and an amorphous, low porous scaling layer was formed on the membrane surface instead of crystallites (Figure 5). As a consequence, a decline of MD process efficiency was larger in the case of antiscalant addition. This phenomenon increased along with increased antiscalant concentration [27]. Moreover, the industrial effluents often contain such large amounts of salts that a dosage of antiscalants would have to be very large, and their efficiency would be doubtful.

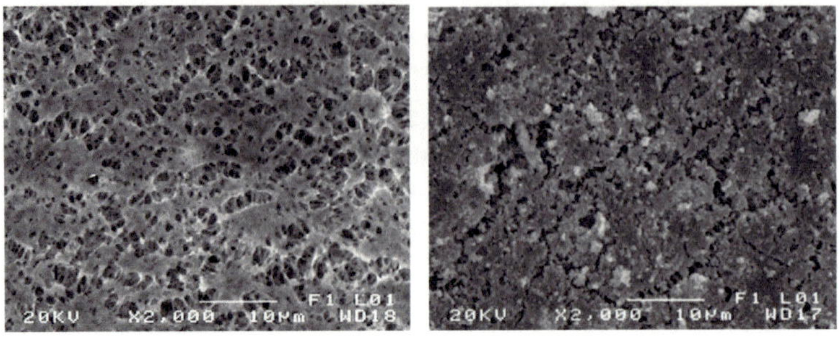

Figure 5. SEM images of Accurel PP S6/2 membranes. A) internal surface of new membrane; B) deposit formed on the membrane surface during MD process of tap water with 10 ppm of $(NaPO_3)_n$.

Effluents desalination by membrane distillation can be restricted by solutes precipitation on the membrane surface, what resulted in a progressive decline of the permeate flux. In such a case there are two solutions: a) the crystals are formed on the membrane surface and they are systematically removed, or b) by using the integrated systems the formation of a scaling layer on the membrane surfaces is restricted, and the crystals formed in the solution subjected to the concentration are separated in the external devices such as net crystallisers [28-30].

In the last case, we stated that the membrane scaling could be limited by the salts precipitation (e.g. $CaCO_3$) in a pre-filtration nets element assembled at the inlet of the MD module (Figure 6).

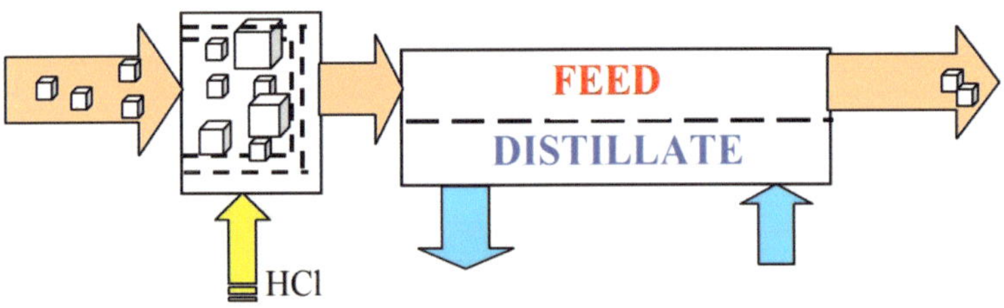

Figure 6. Desalination of effluents using the MD module connected with pre-filtration nets element.

The deposit precipitation (inside the pre filter) eliminated the supersaturation state provided that the induction time of new crystals would be longer than the residue time of feed inside the MD module. The filter efficiency decreased when the deposit layer covered the entire surface of pre-filter, and after a few hours the periodical rinsing of nets by acid solution should be carried-out. The removal of formed deposit (rinsing by HCl solutions) would not result in the membrane wettability.

The above-described solution was used for study of water desalination by MD process. The SEM investigations confirmed that the amount of deposit formed within the interior of MD module was significantly reduced when the filter element was assembled at the module inlet (Figure 7).

http://dx.doi.org/10.5772/49993

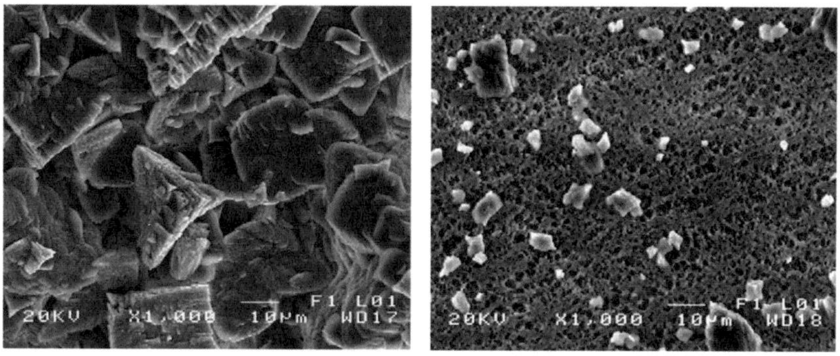

Figure 7. SEM images of membrane surface after 50 h desalination of tap water by MD process. A) MD module without pre-filter nets; B) MD module combined with pre-filter nets element.

The net filter worked as a heterogeneous crystallizer and a large amount of the $CaCO_3$ was deposited on the net surfaces (Figure 8). A periodical removal of the deposit by rinsing with diluted HCl solutions did not cause the membrane wettability [30]. This creates the possibility of application rinsing with acid to restrict a decline of MD installation efficiency during desalination of effluents containing the bicarbonate ions.

Figure 8. SEM image of $CaCO_3$ deposit formed on the nets surface inside the pre-filtration element during MD process of tap water.

The concentration of effluents can be also carried out with a cyclic removal of deposits formed on the membrane surface. However, this should be performed with a sufficient frequency, so that the formed crystals will penetrate the membrane wall in the least degree. In Figure 9 was shown the membrane with deposit precipitated during MD of the geothermal water, subjected to the concentration from 120 to 286 g NaCl/dm^3. The crystallization of calcium sulphate was observed when the concentration of sulphates achieved a level of 2.4-2.6 g $SO_4^{\ 2-}$/dm^3. The SEM-EDS line analysis of membrane cross-section demonstrated that significant scale amounts were found up to the depth of 80-100 μm inside the membrane wall (Figure 10). Therefore, the crystals of $CaSO_4$ can damage the thin walls of capillary membranes. However, the fresh geothermal water dissolved the $CaSO_4$ deposit from the membrane surface. Using a batchwise mode of feeding the MD installation, the concentration of geothermal water was carried out over 800 h, without a significant loss of efficiency of used MD module [31].

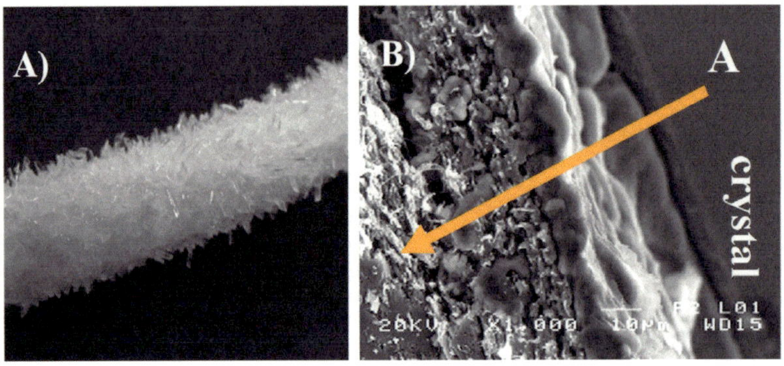

Figure 9. SEM images of $CaSO_4$ deposited on the membrane surface during geothermal water concentration by MD process. A) external surface of capillary Accurel PP V8/2 HF membrane; B) cross-section of membrane with salt deposit.

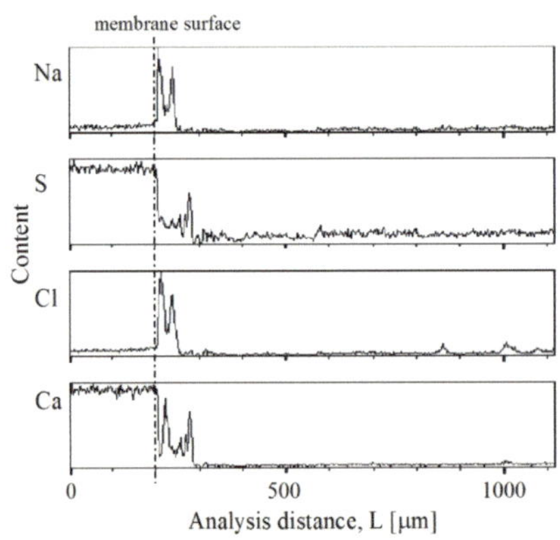

Figure 10. The results of SEM-EDS line analysis (direction A from Figure 9B).

A long-term exploitation of the MD module with a periodical removal of deposits formed on the membrane surface may result in the membrane degradation. The SEM image of the membranes used over a period of several months for water demineralisation was shown in Figure 11. The formed deposit of $CaCO_3$ was removed at every 50-100 h by a periodical rinsing of module with 3-5 wt% HCl [32]. A thickness of damaged membrane did not exceed 50 μm, however, the MD process could be still operate because the total thickness of the membrane amounted 400 μm. Moreover, the obtained results unequivocally demonstrated that the application of methods allowed to restrict the scaling intensity of MD membranes is recommended.

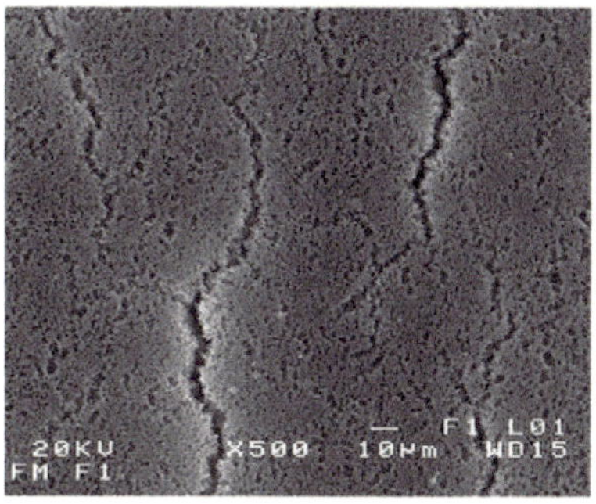

Figure 11. SEM image of the surface of Accurel PP S6/2 membrane degraded as a result of $CaCO_3$ scaling (periodical removal by HCl solution).

A continuous concentration of the saturated NaCl solutions was successfully carried out in a MD installation integrated with the three-stages crystallizer (Figure 12). The amount of salt separated in the crystallizer was varied in the range of 20–102 kg NaCl/m^{2}24h, and was found to be dependent on the feeding solution concentration as well as on the quantity of water removed from the brine in the MD process [14].

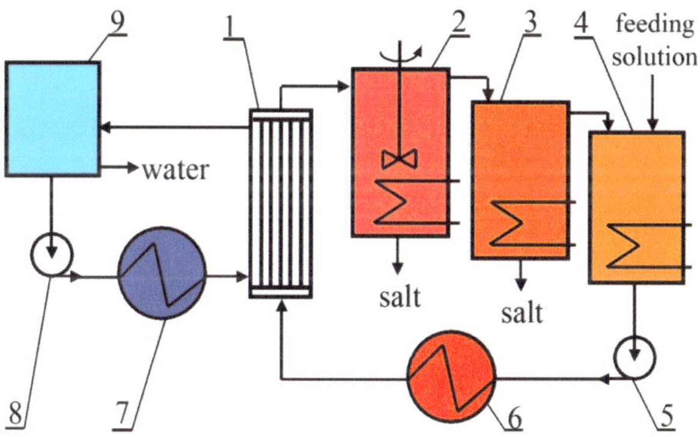

Figure 12. The experimental set-up for DCMD integrated with crystallizer. 1 - MD module, 2, 3, 4 – crystallisation system, 5, 8 - pump, 6 - heating system, 7 - cooling system, 9 - distillate reservoir.

3. Effluents treatment

Water is a solvent commonly used in various technological processes. After the separation of products, the technological operations generate the effluents containing the residues of products and the raw materials. In many cases such effluents after the treatment and supplementation with raw materials could be recycled to the technological process. The accumulation of waste products and salts in these effluents often preclude such solutions. Therefore, the removal of these substances by the methods of selective separation would allow to eliminate this restriction.

The effluents generated during the process of ethanol production constitute the example of such effluents. A significant amount of wastewater discharged from the distillation column (stillage) creates a serious ecological problem. The membrane processes enables the removal of biomass and salts from the post-reaction streams, and such treated effluents can be re-used in the bioreactors. Wastewater poses a threat to the environment and must be managed, which leads to a significant increase in the energy consumption during the process of ethanol production. The energy cost is the second largest factor in ethanol production next to the costs of raw material consumption. The majority of energy is used for broth distillation that contains only 5-12% ethanol after completing the sugar fermentation process [33, 34].

An increase in the ethanol concentration would lower the cost of distillation, but it is difficult to achieve higher concentrations in the classic feed-batch fermentation due to the inhibition phenomena [35, 36]. Another serious problem is associated with large volume of wastewater discharged from the distillation column (stillage). One of the primary methods used in stillage management is the concentration by evaporation and subsequently drying in order to prepare a protein additive for animal feeding. The condensates produced from stillage evaporation process constitute low-polluted water, and can be recycled as dilution water for the fermentation step. The recycling or reuse of low-contaminated wastewater after an appropriate treatment allows to limit its environmental impact. The major drawback is the presence of compounds, such as aliphatic acids (formic, acetic, propionic, butyric, valeric and hexanoic), alcohols (2,3-butanediol), aromatic compounds (phenyl-2-ethyl-alcohol) and furane derivatives (furfural), which inhibit the fermentation [36]. A substantial part of these compounds can be removed by the treatment of condensate with reverse osmosis [36, 37].

The above problems can be successfully solved by reduction of the amounts of stillage, which can be achieved by the application of membrane bioreactors for ethanol production [38, 39]. The membrane separation retains the yeast cells in the bioreactor, which facilitates the distillation of ethanol from the obtained filtrate (Figure 13).

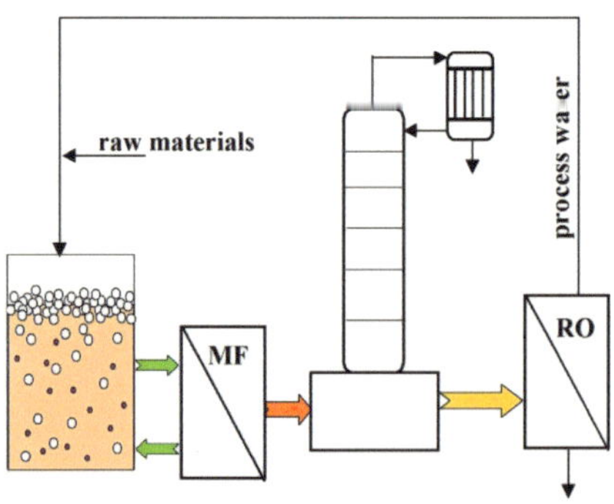

Figure 13. Continuous membrane bioreactor with process water re-use system.

A continuous dosing of the substrates and an increase of the yeast concentration in the broth improves the productivity and efficiency of the bioreactor. Microfiltration (MF) is usually used as the separation method, although it does not reduce the amount of effluents produced during ethanol distillation. However, a clear solution of ethanol without any microorganisms was obtained after the MF process, and the expensive process of stillage evaporation is no longer necessary. After the removal of organic acids e.g. by reverse osmosis [36], such solution can be reused as the technological water in the fermentation process.

Moreover, a continuous fermentation is more attractive than the batch process due to its higher productivity, better process control and improved yields [40-42]. The challenge is how to effectively remove the yeast-produced metabolites from the broth. The use of MD (Figure 14) for the removal of ethanol and other volatile metabolites from broth will both decrease the inhibitory effect of these compounds on microbial culture and reduce the costs of further concentration of alcohol [43].

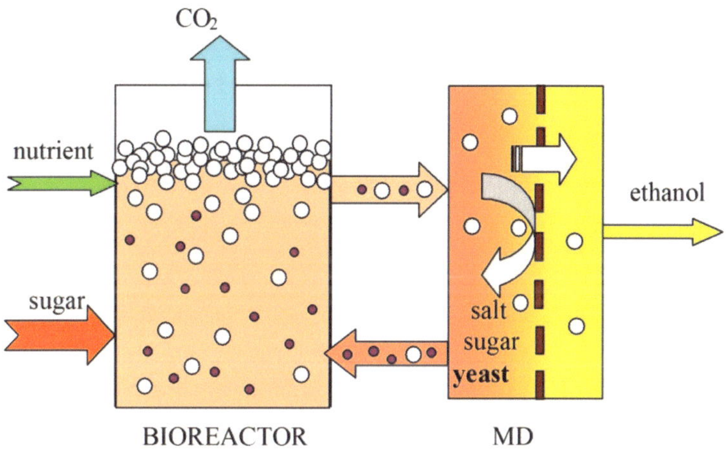

Figure 14. Ethanol production in membrane bioreactor integrated with MD process.

In the case of ethanol production in the MD bioreactor, the concentration of obtained distillate (the enrichment coefficient) is a more important parameter than the magnitudes of permeate flux. The higher the enrichment coefficient, the lower the costs associated with further concentration of the alcohol solution. The results of MD treatment of the fermenting broth were markedly different from the separation of standard solutions of ethanol. During the experiments, the ethanol concentration in the broth varied in the range 20-50 g EtOH/dm^3. The obtained enrichment coefficient amounted to 8-12, and it was two-fold higher than that obtained for the standard solutions of ethanol. It was found that bubbles of the CO_2 formed during the fermentation had a significant influence on the results [44]. Due to the presence of the bubbles forming a layer adjacent to the membrane, the layer is enriched in alcohol, which affects the separation result during MD process.

The main advantage of MD over conventional distillation processes is that the membrane distillation takes place at a temperature below the normal boiling point of broth solutions. The major requirement of MD process is that the used membranes must not be wetted by treated solutions. The performed studies demonstrated that the broth subjected to the separation did not affect the hydrophobic properties of the polypropylene membrane assembled in the MD modules. Moreover, the MD process was successfully applied for a long-term study to remove the volatile components from the fermentation broth. Besides ethanol, propionic and acetic acids were transferred from the broth to the distillate [43, 44]. Therefore, the course of the fermentation carried out in the membrane distillation bioreactor considerably accelerates the fermentation rate and increases the efficiency by a selective removal of the fermentation products.

A selective separation of ethanol only has the effect of increasing the concentration of other metabolites and substrates that have not been used. It creates an unfavourable reaction environment for yeast and, as a result, suddenly decreases the number of yeast cells and fermentation efficiency after a few dozen hours [45]. In such case, besides a selective removal of separated volatile metabolites a part of broth (bleeding method) should be also removed from the bioreactor. The collected solution can be subjected to the classical distillation (ethanol removal), and the obtained stillage can be further treated with the use of MF process followed by NF. The NF permeate can be recycled into the bioreactor (Figure 15).

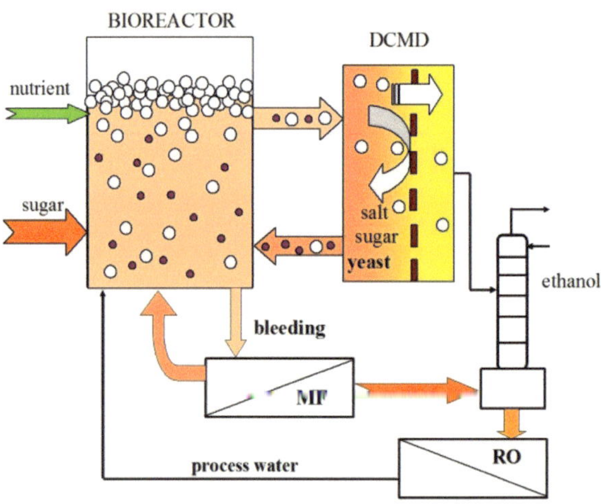

Figure 15. Integrated membrane system for continuous ethanol production with effluents treatment.

The MF is often utilized for the purification and sterilization of various biological solutions, including fermentation broths [46]. However, fouling causes a decrease of permeate flux resulting from growing resistance of membrane, hence, a deterioration of efficiency of processes and finally, shortening the membrane life (Figure 16).

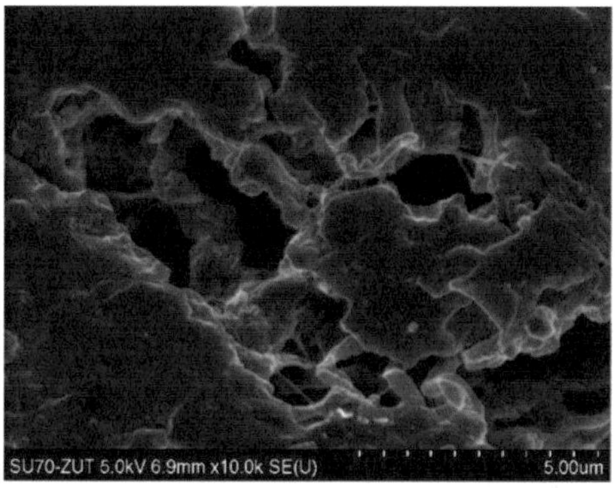

Figure 16. SEM image of deposit formed on the membrane surface during the MF process of broth.

It was demonstrated that when the filtered effluents contains only the particles larger than the membrane pores, the main mechanisms of fouling is a pore clogging and the formation of filter cake [47]. The removal of fouling deposits requires the chemical cleaning, the objective of which is to restore the permeability and selectivity of the membranes process while preserving the hygienic conditions of operation [48]. The used cleaning agents clean the membrane through the removal of foulants, change in their morphology or a modification of properties of fouling layer (Figure 17). It was found that the effectiveness of chemical cleaning depends on several factors including: temperature, pH, cleaning agents concentration, residence time of cleaning solution in a module and the operating conditions such as cross-flow velocity and the transmembrane pressure [49].

Figure 17. SEM image of membrane surface after chemical cleaning of MF module.

The literature review reveals that it is more advantageous to use the ceramic membranes instead of the polymeric membranes, because they exhibit high mechanical and chemical resistance [50]. This enables the operation of ceramic module under high pressures and the application of backflush, and offers the resistance on extremely high temperatures and concentrations [50, 51].

The MF process of yeast suspensions was carried out using a single-channel ceramic membrane and the polypropylene membrane with similar dimensions of the pores (0.2 μm). The changes of the values of relative fluxes during the broth separation were shown in Figure 18.

In the case of MF with the use polymeric membrane, a decline of the permeate flux was a slightly lower. After completing the MF process studies, the determination of decline of maximum permeates flux was performed. It was found, that the maximal permeate flux for the ceramic membrane decreased to 30% of the initial value after performing MF of yeast suspension for 120 min. This flux decline was larger by about 10% in comparison with that obtained for the polymeric membrane. After completing the microfiltration process, the installation was rinsed with water several times, what allowed to remove the majority of suspensions from the installation. Subsequently, the permeate flux for clean water was checked in order to determine the actual efficiency of the membranes. This procedure allowed to determine a decline of maximal permeates flux after completed stage of the studies. The maximum permeate flux for the ceramic membrane decreased to 40% of the initial value after performing the microfiltration of yeast suspension for 180 min. This flux decline was larger by about 20% in comparison with that for the polymeric membrane.

In order to compare the effectiveness of used chemical cleaning of both membranes, the same cleaning procedure was used: rinsing the installation with water, rinsing with 1% solution of sodium hydroxide for 20 min followed by rinsing with 0.5% solution of orthophosphoric acid (20 min). The application of chemical cleaning enabled the restoration of the initial efficiency for the polymeric membrane almost in 100%, contrary to that was observed in the case of ceramic membrane (Figure 19). Based on this fact it was concluded, that the ceramic membrane requires a long cleaning process, what affects adversely the costs of membrane process with the use of ceramic membranes.

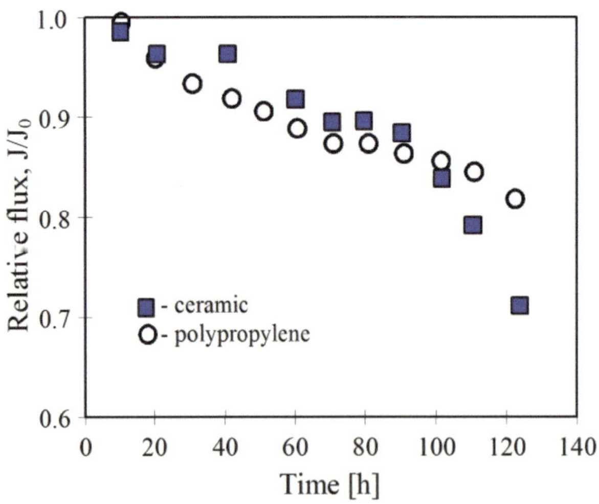

Figure 18. Changes of the relative fluxes obtained during MF of yeast solution (3 g/L) with ceramic and polypropylene membranes.

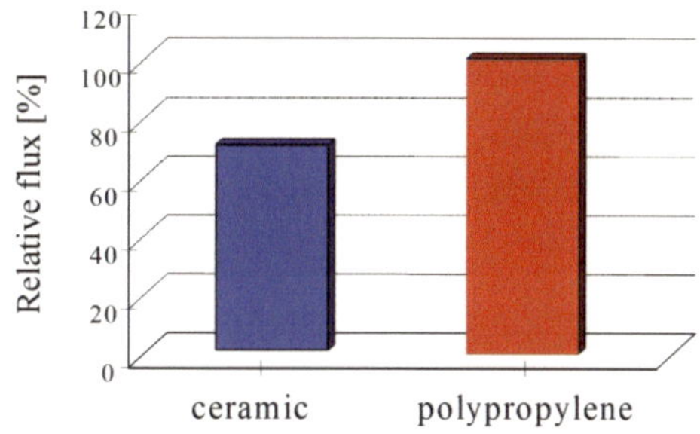

Figure 19. The results of chemical cleaning of polypropylene and ceramic membrane.

The chemical cleaning of ceramic membrane was repeated 4 times, however, the expected results were not achieved. Further studies demonstrated that after the separation of biological solutions the highest effectiveness of ceramic membrane cleaning was achieved as a result of using multi-stage cleaning procedure involving chemical cleaning – alkaline and acidic and back flushing.

4. Conclusions

The industrial effluents can contain small amounts of impurities, hence, their recirculation to the technological process is hindered. With regard to this, these effluents constitute a wastewater, which are often dangerous to the environment. In many cases, solutes (e.g. NaCl) contained in these effluents are not poisonous, however, their excessive amount renders their disposal into the environment impossible. The removal of these components can transform the industrial effluents from waste into the reusable raw materials.

The membrane processes, especially in the form of integrated systems, allows to treat even a complex effluents. Moreover, their application allows to modify the technology, and as a result, the amount of generated wastewater will be significantly reduced.

Acknowledgements

The Polish State Committee for Scientific Research is acknowledged for the support of this work (2009–2012).

Author details

Marek Gryta*

Address all correspondence to: marek.gryta@zut.edu.pl

West Pomeranian University of Technology, Szczecin, Poland

References

[1] Drioli, E., & Romano, M. (2001). Progress and new perspectives on integrated membrane operations for sustainable industrial growth. Industrial & Engineering Chemistry Research, 2001;40(5) , 1277-1300.

[2] Singh, R. (2006). Hybrid membrane systems for water purification. Oxford: Elsevier.

[3] Gűnder, B. (2001). The membrane-coupled activated sludge process in municipal wastewater treatment. Lancaster: Technomic.

[4] Schorr, M. editor. (2011). Desalination, trends and technologies. Rijeka: InTech.

[5] Nunes, S. P., & Peinemann, K. V. editors. (2006). Membrane technology in the chemical industry. Weinheim: Wiley; 2006.

[6] Gryta, M., Tomaszewska, M., Morawski, A. W., & Grzechulska, J. (2001). Membrane distillation of NaCl solution containing natural organic matter. *Journal of Membrane Science*, 181(2), 279-287.

[7] Curcio, E., Ji, X., Di Profio, G., Sulaiman, A. O., Fontananova, E., & Drioli, E. (2010). Membrane distillation operated at high seawater concentration factors: Role of the membrane on $CaCO_3$ scaling in presence of humic acid. *Journal of Membrane Science*, 346(2), 263-269.

[8] Gryta, M. (2002). Direct contact membrane distillation with crystallization applied to NaCl solutions. *Chemical Papers*, 56(1), 14-19.

[9] Teoh, M. M., Bonyadi, S., & Chung, T. S. (2008). Investigation of different hollow fiber module designs or flux enhancement in the membrane distillation process. Journal of Membrane Science, 2008;311(1-2) , 371-379.

[10] Mericq, J. P., Laborie, S., & Cabassud, C. (2010). Vacuum membrane distillation of seawater reverse osmosis brines. *Water Research*, 44(18), 5260-5273.

[11] El-Bourawi, M. S., Ding, Z., Ma, R., & Khayet, M. (2006). A framework for better understanding membrane distillation separation process. Journal of Membrane Science, 2006;285(1-2) , 4-29.

[12] Gryta, M., Karakulski, K., Tomaszewska, M., & Morawski, W. (2006). Demineralization of water using a combination of MD and NF(RO). Desalination, 2006;200(1-3) 451-452.

[13] Karakulski, K., & Gryta, M. (2005). Water demineralisation by NF/MD integrated processes. Desalination, 2005;177(1-3) , 109-119.

[14] Gryta, M. (2002). Concentration of NaCl solution by membrane distillation integrated with crystallisation. Separation and Purification Technology , 37(15), 3535-3558.

[15] Karakulski, K., Gryta, M., & Sasim, M. (2006). Production of process water using integrated membrane processes. *Chemical Papers*, 60(6), 416-421.

[16] Gryta, M., Karakulski, K., Tomaszewska, M., & Morawski, A. (2005). Treatment of effluents from the regeneration of ion exchangers using the MD process. Desalination, 2005;180(1-3) , 173-180.

[17] Qu, D., Wang, J., Wang, L., Hou, D., Luan, Z., & Wang, B. (2009). Integration of accelerated precipitation softening with membrane distillation for high-recovery desalination of primary reverse osmosis concentrate. *Separation and Purification Technology*, 67(1), 21-25.

[18] Tomaszewska, M., Gryta, M., & Morawski, A. W. (2001). Recovery of hydrochloric acid from metal pickling solutions by membrane distillation. Separation and Purification Technology, 2001; 22-23 , 591-600.

[19] Oluji, Ž., Behrens, M., Sun, L., & Fakhri de Graauw, J. (2010). Augmenting distillation by using membrane based vapor-liquid contactors: An engineering view from Delft. Journal of Membrane Science, 2010;350(1-2) 19-31.

[20] Macedonio, F., & Drioli, E. (2008). Pressure-driven membrane operations and membrane distillation technology integration for water purification. Desalination, 2008;223(1-3) , 396-409.

[21] Gryta, M. (2009). $CaSO_4$ scaling in membrane distillation process. *Chemical Papers*, 63(2), 146-151.

[22] Sheikholeslami, R. (2007). Fouling in membranes and thermal units. L'Aquila: Balaban Desalination Publications.

[23] Gryta, M. (2008). Alkaline scaling in the membrane distillation process. Desalination, 2008;228(1-3) , 128-134.

[24] Gryta, M. (2008). Fouling in direct contact membrane distillation. *Journal of Membrane Science*, 325(1), 383-394.

[25] Gryta, M. (2010). Application of membrane distillation process for tap water purification. *Membrane Water Treatment*, 1(1), 1-12.

[26] Gryta, M. (2008). Chemical pretreatment of feed water for membrane distillation. *Chemical Papers*, 62(1), 100-105.

[27] Gryta, M. (2012). Polyphosphates used for membrane scaling inhibition during water desalination by membrane distillation. *Desalination*, 285(1), 170-176.

[28] Gryta, M. (2010). Desalination of thermally softened water by membrane distillation process. Desalination, 2010;257(1-3) , 30-35.

[29] Gryta, M. (2011). The influence of magnetic water treatment on $CaCO_3$ scale formation in membrane distillation process. *Separation and Purification Technology*, 80(2), 293-299.

[30] Gryta, M. (2009). Scaling diminution by heterogeneous crystallization in a filtration element integrated with membrane distillation module. *Polish Journal of Chemical Technology*, 11(2), 60-65.

[31] Gryta, M., & Palczyński, M. (2011). Desalination of geothermal water by membrane distillation. Membrane Water Treatment , 2(3), 147-158.

[32] Gryta, M. (2005). Long-term performance of membrane distillation process. Journal of Membrane Science, 2005;265(1-2) , 153-159.

[33] Demirbas, A. (2007). Progress and recent trends in biofuels. Progress in Energy and Combustion Science, 2007;33 , 1-18.

[34] Bai, F. W., Anderson, W. A., & Moo-Young, M. (2008). Ethanol fermentation technologies from sugar and starch feedstocks. *Biotechnology Advances*, 26(1), 89-105.

[35] Sassner, P., Galbe, M., & Zacchi, G. (2008). Techno-economic evaluation of bioethanol production from three different lignocellulosic materials. *Biomass and Bioenergy*, 32(5), 422-430.

[36] Morin-Couallier, E., Payot, L. T., Pastore, Bertin. A., & Lameloise, M. L. (2006). Recycling of distillery effluents in alcoholic fermentation. *Applied Biochemistry and Biotechnology*, 133(3), 217-238.

[37] Morin-Couallier, E., Salgado-Ruiz, B., Lameloise, M. L., & Decloux, M. (2006). Usefulness of reverse osmosis in the treatment of condensates arising from the concentration of distillery vinasses. Desalination, 2006;196(1-3) , 306-317.

[38] Takaya, M., Matsumoto, N., & Yanase, H. (2002). Characterization of membrane bioreactor for dry wine production. *Journal of Bioscience and Bioengineering*, 93(2), 240-244.

[39] Park, B. G., Lee, W. G., Chang, Y. K., & Chang, H. N. (1999). Long-term operation of continuous high cell density culture of Saccharomyces cerevisiae with membrane filtration and on-line cell concentration monitoring. *Bioprocess Engineering*, 21(2), 97-100.

[40] Choi, G.-W, Kang, H.-W., & Moon, S.-K. (2009). Repeated-batch fermentation using flocculent hybrid, Saccharomyces cerevisiae CHFY0321 for efficient production of bioethanol. *Applied Microbiology and Biotechnology*, 84(2), 261-269.

[41] Kargupta, K., Datta, S., & Sanyal, S. K. (1998). Analysis of the performance of a continuous membrane bioreactor with cell recycling during ethanol fermentation. *Biochemical Engineering Journal*, 1(1), 31-37.

[42] Park, B. G., Lee, W. G., Chang, Y. K., & Chang, H. N. (1999). Long-term operation of continuous high cell density culture of Saccharomyces cerevisiae with membrane filtration and on-line cell concentration monitoring. *Bioprocess Engineering*, 21(2), 97-100.

[43] Barancewicz, M., & Gryta, M. (2012). Ethanol production in a bioreactor with an integrated membrane distillation module. *Chemical Papers*, 66(2), 85-91.

[44] Gryta, M. (2001). The fermentation process integrated with membrane distillation. Separation and Purification Technology, 2001;24(1-2) , 283-296.

[45] Gyamerah, M., & Glover, J. (1996). Production of ethanol by continuous fermentation and liquid-liquid extraction. *Journal of Chemical Technology and Biotechnology*, 66(2), 145-152.

[46] Sondholi, R., & Bhave, R. (2001). Role of backpulsing in fouling minimization in crossflow filtration with ceramic membranes. Journal of Membrane Science , 186(1), 41-52.

[47] Stopka, J., Bugan, S. G., Broussous, L., Schlosser, S., & Larbot, A. (2001). Microfiltration of yeast suspensions through stamped ceramic membranes. Separation and Purification Technology, 2001;25(1-3) , 535-543.

[48] Bird, M. R., & Bartlett, M. (2002). Measuring and modelling flux recovery during the chemical cleaning of MF membranes for the processing of whey protein concentrate. *Journal of Food Engineering*, 53(2), 143-152.

[49] Blanpain-Avet, P., Migdal, J. F., & Benezech, T. (2009). Chemical cleaning of a tubular ceramic microfiltration membrane fouled with a whey protein concentrate suspension-Characterization of hydraulic and chemical cleanliness. Journal of Membrane Science, 2009;337(1-2) , 153-174.

[50] Hofs, B., Ogier, J., Vries, D., Beerendonk, E. F., & Cornelissen, E. R. (2011). Comparison of ceramic and polymeric membrane permeability and fouling using surface water. *Separation and Purification Technology*, 79(3), 365-374.

[51] Kim, J., & Van Der Bruggen, B. (2010). The use of nanoparticles in polymeric and ceramic membrane structure: Review of manufacturing procedures and performance improvement for water treatment. *Environmental Pollution*, 158(7), 2335-2349.

Chapter 3

Novel Biopolymer Composite Membrane Involved with Selective Mass Transfer and Excellent Water Permeability

Peng Wu and Masanao Imai

Additional information is available at the end of the chapter

http://dx.doi.org/10.5772/50697

1. Introduction

Applications of bio-polymeric materials have increased significantly for both textile engineering and medical sciences. Use of biopolymer in place of artificial polymers has been increasing due to stringent environmental regulations [1]. The development of new-generation materials that extend the industrial and biomedical applications of membrane processes will require a high level of control of the characteristics of the base polymeric support layer [2]. Current research in membrane science is now focusing more on biopolymers from natural raw material with well-defined structure to develop new membrane materials [3]. Noticeably, biopolymer production can be sustainable, carbon neutral, and renewable because biopolymers are made from sea or land plant materials that can be grown year after year indefinitely. Novel biopolymer membranes enable separation based on other driving forces like electrical charge and physicochemical interactions, and with appropriate functional groups can provide applications such as tunable water permeation and separation, toxic metal capture, toxic organic dechlorination, and biocatalysts [4-5].

Biopolymers for stabilizers, thickeners, and gelling agents were extracted from various raw natural resources. They determine a number of critical functions including moisture binding, control, structure, and flow behavior that enable organisms to thrive in a natural environment [6]. A number of the typical biopolymers from natural resources such as alginate, cellulose, and chin/chitosan have been applied for functional polymer networks (e.g., carriers for controlled drug release, membranes with regulated permeability, sensor devices, and artificial muscles). For these purposes, proper responses to changes in external physicochemical conditions and developed internal microstructure of the gels are required. Interest in the behavior of biopolymer gels and networks has grown significantly. The various hydrophilic bio and/or artificial polymers that can be used for membrane formation are also discussed.

Various novel membrane materials and systems have been developed and applied. The technological benefits of such membrane materials and systems have begun to be identified for a wide range of applications for controlled drug delivery, chemical separation, water treatment, bio-separation, chemical sensors, tissue engineering, etc. There have been two main subjects of research in the field of biopolymer membrane materials and systems: development of novel and efficient biopolymer materials and improvement of capability of membrane processes and operations [7].

This chapter discusses the novel function of polysaccharides (κ-carrageenan (κC) and pullulan (P)) in membrane formation and molecular-size screening. The κ-carrageenan mass fraction (F_C) was a key factor in determining membrane characteristics for both selected molecular permeability and mechanical strength.

2. Development of Biopolymer membranes

When discussing biopolymer gelation, the biopolymer types of interest fall naturally into two categories: protein and polysaccharides. A second classification is in terms of the molecular networks underlying the gels, that is, in terms of associative and particulate networks [8]. The present status of biological and ecological research demands much more emphasis on efficient biopolymers with multiple applications such as membrane-separation engineering.

One of the most common membrane types currently in use is the asymmetric cellulose acetate (CA) membrane. This high-flux, high-rejection membrane was developed in the early 1960s by Loeb and Sourirajan [9]. Chitin, poly (β-(1-4)-N-acetyl-D-glucosamine), is a natural polysaccharide of major importance, first identified in 1884. When the degree of deacetylation of chitin reaches about 50% (depending on the origin of polymer), it becomes soluble in aqueous acidic media and is called chitosan [10-11]. Chitosan membranes have been explored in many uses, such as in water–ethanol pervaporation [12-14], enzyme immobilization and cationic specimen transportation [15-16], protein separation [17] and concentration, controlled ingredient-release, and environmental applications [18-19]. Among the various biopolymers, alginate is the most studied matrix for membrane separation technology [20]. Hirst and Rees (1965) were the first to postulate that alginate is a polymer of mannuronic acid and guluronic acid having 1,4 linkage. Kashima and Imai (2011) investigated the α-L-guluronic acid chain with regard to regulation of the mass-transfer characteristics of the alginate membrane [21]. Many other biopolymers also consist of membrane structures. Exploiting and improving the chemical and mechanical properties of biopolymer membranes will create many more applications in the membrane industry.

3. Novel biopolymer membrane materials (κ-carrageenan & pullulan)

Biopolymers of marine algae origin are ubiquitous in surface waters and have attractive potential. The seaweed extractives of commercial importance fall into three main groups, two of which (agar and carrageenans) are derived from red algae, and the third (alginates) from brown algae. All three types of extractive are associated with the cell walls of the algae and resemble cellulose in basic molecular organization. Red algae are considered as the most important resource of many biologically active metabolites compared to other algal classes [22-23]. Marine algae as carrageenans have optimum growth conditions with sufficient sun light, stable temperature, and no climate change impact on the ground; stable harvests are thus expected. Marine algae can be produced in virtually unlimited amounts around seafaring nations. It contributes noticeably on the preventing Global Warming and coexistent with fishery.

Carrageenans are large, highly flexible molecules that curl and form helical structures. They are widely used in food and other industries as thickening and stabilizing agents. Carrageenans consist of alternating copolymers of α-(1→3)-D-galactose and β-(1→4)-3,6-anhydro-D- or L-galactose. Several isomers of carrageenan are known (κ-, ι-, and λ-carrageenans), and they differ in the number and position of the ester sulfate groups on the repeating galactose units. κ-Carrageenan has only one negative charge per disaccharide and tends to form a strong and rigid gel. The gelling power of κ-carrageenans imparts excellent film-forming properties, and κ-carrageenan forms a firm gel with the aid of potassium ions. Hot solutions of κ-carrageenans set when cooled below the gel point, which is between 30° and 70°, depending on the cations and other ingredients present, to form a range of gel textures. The two-step gel mechanism is illustrated in Fig. 1, with stage B being elastic (iota) and stage C being brittle (kappa).

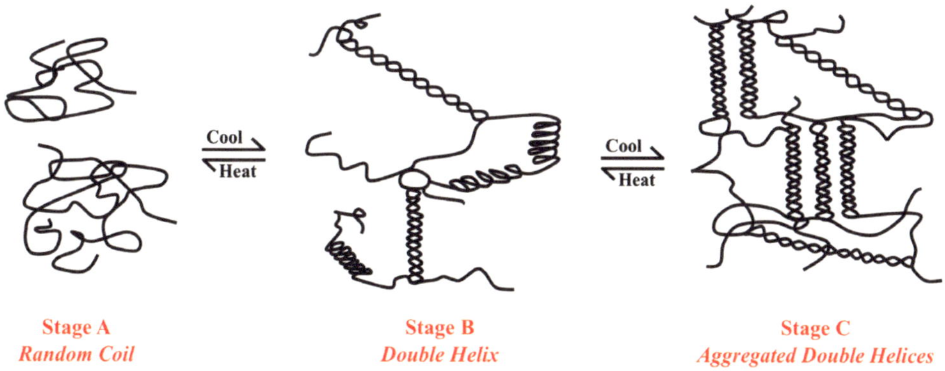

Figure 1. Models of conformational transition of κ-carrageenan and ι-carrageenan.

κ-Carrageenan selects potassium ions to stabilize the junction zones within the characteristically firm, brittle gel. Potassium ions counter sulphate charges without sterically hindering close approach and double-helix formation (Fig. 2) [24-30].

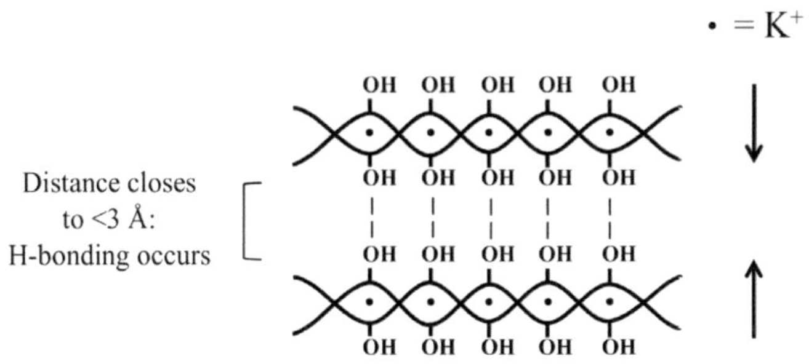

Figure 2. The gelation mechanism of κ-Carrageenan crosslinked by K^+ ions.

Pullulan is an extracellular glucan elaborated by a fungus of the genus Aureobasidium, commonly called black yeast. The structure of pullulan is a linear glucan consisting of repeating units of maltotriose joined by α-D-(1→6) linkages. The safety of pullulan in foods is supported by its chemical composition, the purity of the final product, a series of toxicological studies, and the fact that it has been used about 30 years as an ingredient in human foods in Japan [31-33]. Recently, the demand for pullulan has rapidly increased for films and hard capsules, and its use in these fields is expected to grow [34-35]. The major interest in pullulan concerns its capacity to form strong, resilient films and fibers [36]. Pullulan can be used on its own or combined with other thickeners or gelling agents. The stringiness of pullulan may be a disadvantage for some applications, but this can be modified by adding a small amount of another polysaccharide such as carrageenan or xanthan gum [37]. Combinations of κ-carrageenan and pullulan achieve gel-network strengths and elasticity between the two extremes and consistent with the ratio used (Fig. 3).

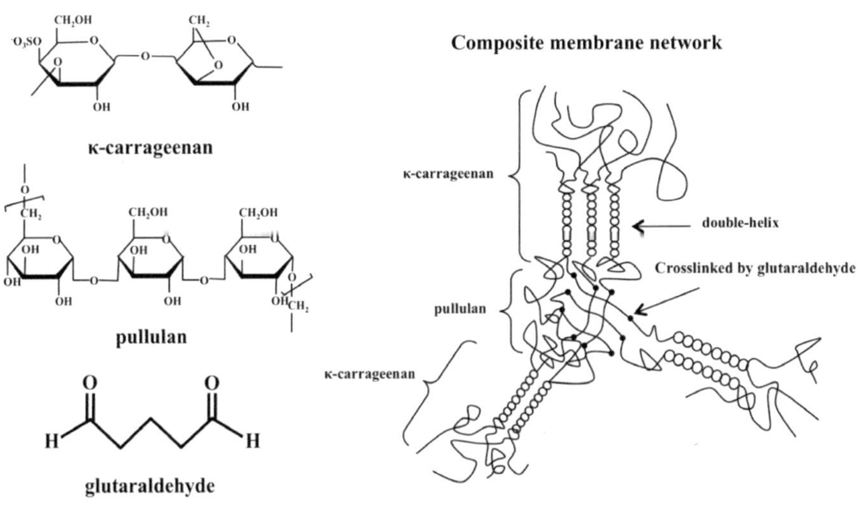

Figure 3. Schematic representation of composite κ-carrageenan-pullulan chains.

http://dx.doi.org/10.5772/50697

In our study, original biopolymer composite membrane was successfully prepared from marine biopolymer (κ-carrageenan) and food polysaccharide (pullulan). Selective mass transfer and excellent water permeability were achieved. The membrane was characterized from the mass fraction of κ-carrageenan. The attractive potential of marine biopolymer (κ-carrageenan) combined with polysaccharide (pullulan) was demonstrated in membrane-separation engineering. The authors focused on the complex cross-linked biopolymers (κ-carrageenan and pullulan) regulating mass-transfer flux. The membrane was prepared by a simple casting method. The κ-carrageenan-pullulan composite membrane has sufficient mechanical strength for practical use and excellent mass-transfer characteristics, especially for molecular-size screening.

4. Polymer membranes preparation

The most important part in any membrane separation process is choosing the membrane material. Membranes have very different structures, functions, transport properties, transport mechanisms, and materials. The methods of making membranes are just as diverse as the membranes are. The methods of making membranes are considering the large diversity suited for technical application. The following characteristic of membranes determine separation capability.

- Membrane materials.

Organic polymers, inorganic materials (oxides, ceramics, and metals), or composite materials.

- Membrane cross-section.

Isotropic (symmetric), integrally anisotropic (asymmetric), bi- or multilayer, thin-layer or mixed matrix composites.

- Preparation methods.

Phase separation (phase inversion) of polymers, sol–gel process, interface reaction, stretching, extrusion, track-etching, and micro-fabrication.

- Membrane shape.

Sheet, hollow fiber, capsule.

The process for forming a biopolymer membrane comprises three steps.:

I. Mixing a biomaterial in a solvent to define a gel.

II. Drying the gel to define a sponge having a solvent content.

III. Adjusting the solvent content of the sponge so that the sponge is substantially filled with the solvent.

4.1. Preparing of κ-carrageenan/pullulan composite membranes

κ-Carrageenan and pullulan powders were dissolved in distilled water (70°C) using a magnetic stirrer to prepare film-forming solutions of various blend-weight ratios. All polymer solutions were prepared based on 3g total polymer weight dissolved in 97g of distilled water at 70°C for one hour. In addition, each solution was stirred for one hour at 70°C. Glutaraldehyde solution (30 ~ 130mM) was introduced into the polymer solutions [39]. And then twenty grams of the polymer solutions was then cast into a petri dish, followed by drying in an electrical blast-drying chest at 65°C for 24 hours. The dried membranes (attached to the petri dishes) were immersed in potassium chloride solution [40] (0.1 to 1.0M) for 24 hours. The swollen membrane spontaneously peeled from the petri dish at 25 ± 1°C and was washed clean with pure water for further testing. Membrane samples were tested in triplicate. Pure pullulan single component membrane (cross-linked by glutaraldehyde) was too weak to make a flat membrane in our study [41].

Mass fraction of κ-carrageenan (F_c)	F_c [-]	κ-carrageenan [g]	Pullulan [g]
$F_c = \frac{\kappa - \text{carrageenan}[g]}{\kappa - \text{carrageenan}[g] + pullulan[g]}$	0.33	1.00	2.00
	0.42	1.25	1.75
	0.5	1.50	1.50
	0.58	1.75	1.25
	0.66	2.00	1.00
	0.75	2.25	0.75

Table 1. Mass fraction of κ-carrageenan (F_C).

5. Measurement of biopolymer composite membranes' properties

Commercial membrane applications focus much effort on desalination requirements [42-43], membrane-fouling characterization [44-45], drinking-water disinfection [46-47], industrial waste treatment [48-49], food industry material separation [50-51], adsorption desalination [52-53], biofiltration [54], membrane bioreactor [55-56], thermal distillation [57], electrodialysis desalination [58], reverse-osmosis desalination [59], oil–water separation applications [60], and future membrane and desalination developments. The stress-strain correlation of biopolymer membrane is affected by the origin of polymers, molecular weight, and methods of membrane preparation, conditioning, and cross-linking. Biopolymer membranes may be amorphous homopolymers or heterogeneous, depending on whether they are prepared from a single polymer or from blended polymers [61]. However, the properties of biopolymer membranes are inconsistent with the requirements of industrial-processing technologies, since the range of biopolymers suitable for membrane-separation processes is limited. To expand the application area of commercial membranes, research on improving their properties is necessary.

http://dx.doi.org/10.5772/50697

5.1. Mechanical properties of κ-carrageenan/pullulan composite membrane

A rheometer (CR-DX500, Sun Scientific Co., Ltd., Tokyo, Japan) was used to determine the tensile strength and the percentage elongation at break. Three rectangular-strip specimens (10mm wide, 40mm long) were cut from each membrane for tensile testing. The initial grip separation was set to 20mm, and the crosshead speed was set to 1mm/s. The initial membrane thickness was measured using a micrometer (Mitutoyo, Kanagawa, Japan). The average thickness of the membrane strip was used to estimate the initial cross-sectional area of the membrane sample. Maximum Stress (σ) (MPa) was calculated by dividing the maximum load (N) by the initial cross-sectional area (m^2):

$$\sigma = \frac{T}{(b \times d)} \quad (1)$$

where T is the maximum load (N), b is the width of sample (m), and d is the membrane thickness (m).

Maximum Strain (λ) (%) was calculated as follows:

$$\lambda = \frac{(L - L_0)}{(L_0)} \times 100\% \quad (2)$$

where L_0 is the sample length before deformation and L is the sample length at break [62].

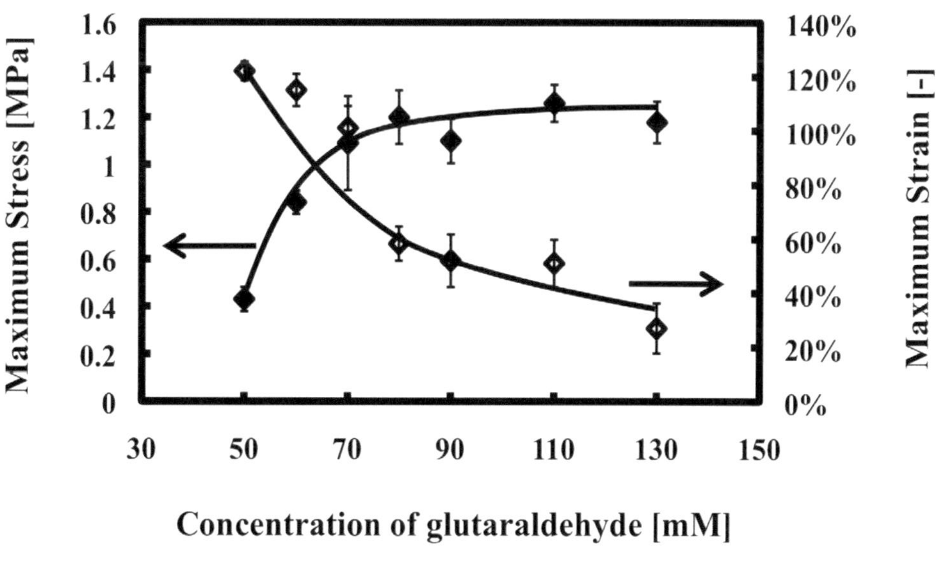

Figure 4. Effect of additive glutaraldehyde concentration on the maximum stress and strain of composite membrane. F_C was set at 0.33.

Dehydration of hydroxyl groups in the polysaccharide chain by glutaraldehyde facilitates formation of polymer networks. The polysaccharide composite membrane was further cross-linked with glutaraldehyde to reduce swelling and increase the structural strength of the membrane as well as to improve its thermal and mechanical stability [63]. The lower F_C membrane is convenient for investigating the influence of dehydration by glutaraldehyde because the lower F_C membrane contains many hydroxyl groups bonding to pullulan. The mechanical stress increased with increasing concentration of glutaraldehyde and became constant over 70mM. Figure 4 presents the polymeric framework of the membrane densely populated with increasing glutaraldehyde concentration.

F_C 0.75 membrane accounts for the largest mass fraction of κ-carrageenan. κ-carrageenan is a key component for constructing the gel structure and for characterizing mechanical strength.

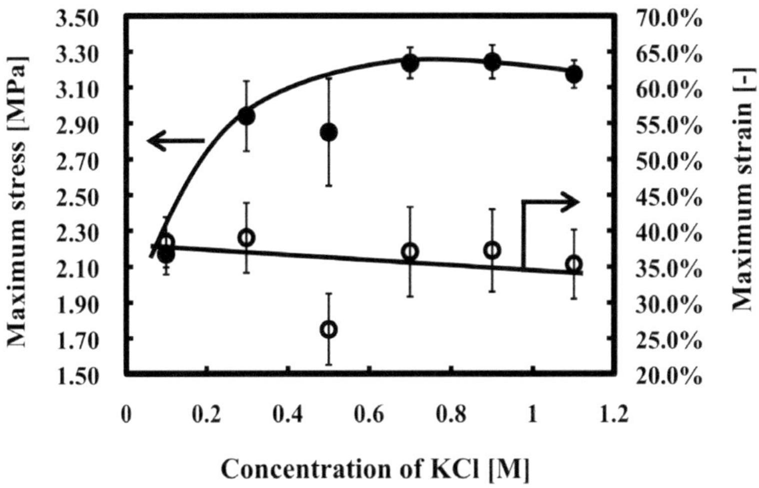

Figure 5. Effect of potassium chloride-immersion on the maximum stress and the strain of composite membrane. F_C was set at 0.75.

The authors prepared κ-carrageenan/pullulan membranes with 90mM of glutaraldehyde added and 0.7M potassium chloride-immersion (Fig. 5)

5.2. Water content

Water content is important for evaluating hydrophilic characteristics. The volumetric water content of the membrane indicated voids in the network that affect the water permeability [64]. Gravimetric methods were used to determine the mass-based water content (W_t) [65]. The water content (W_t) was measured as follows. Membranes were immersed in distilled water at 25±1°C for 1 day to achieve natural hydration and swelling. The membranes were removed from the water bath, and excess water on the membrane surface was removed by filter paper. The mass of the swollen membrane W_w was then determined.

http://dx.doi.org/10.5772/50697

$$W_t = \frac{(W_w - W_d)}{W_w} \times 100\% \quad (3)$$

Here, W_d is the mass of the dried membrane.

5.3. Scanning Electron Microscopy (SEM)

The membranes were snap-frozen in liquid nitrogen then dried in a vacuum freeze dryer (RLE-103, Kyowa Vacuum Engineering. Co., Ltd., Tokyo, Japan) (298 K) for 24 hours. The membranes were then sputter-coated with a thin film of Pt, using a sputter-coater (E-1010 Ion Sputter, Hitachi, Ltd., Tokyo, Japan). Images of cross sections of the membranes were obtained using a scanning electron microscope (Miniscope TM-1000, Hitachi, Ltd.,).

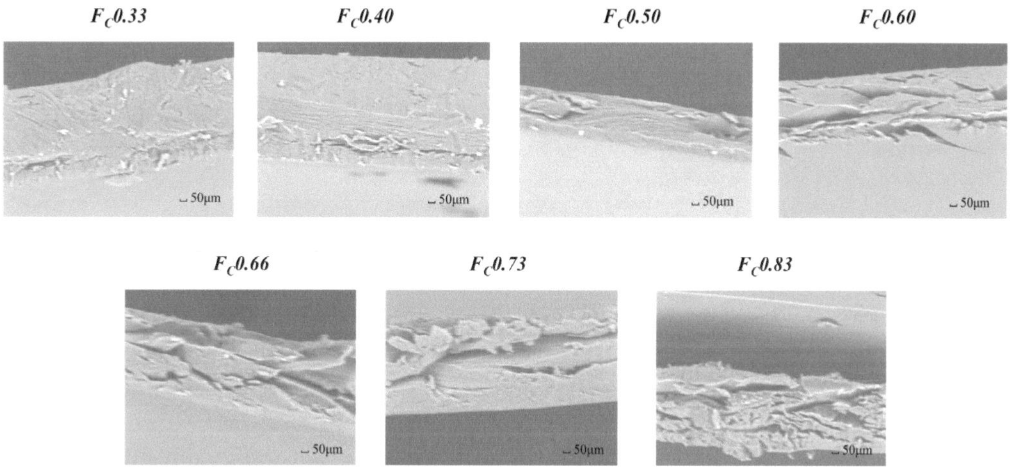

Figure 6. SEM images of κ-carrageenan/pullulan composite membrane.

6. Mass transfer in biopolymer membrane

The prolific application of membrane separation processes in industry today is primarily due to innovations in membrane materials technology. Loeb and Sourirajan (1963) pioneered the first reverse-osmosis (RO) asymmetric cellulose acetate (CA) membrane capable of withstanding the rigors of industrial use [66]. Since then, many types of biopolymer membranes have been developed and commercialized: membranes for microfiltration (MF) [67-68], ultrafiltration (UF) [69-70], nanofiltration (NF) [71-72], gas separation [73-74], and so on. These membrane separation systems are illustrated in Fig. 7.

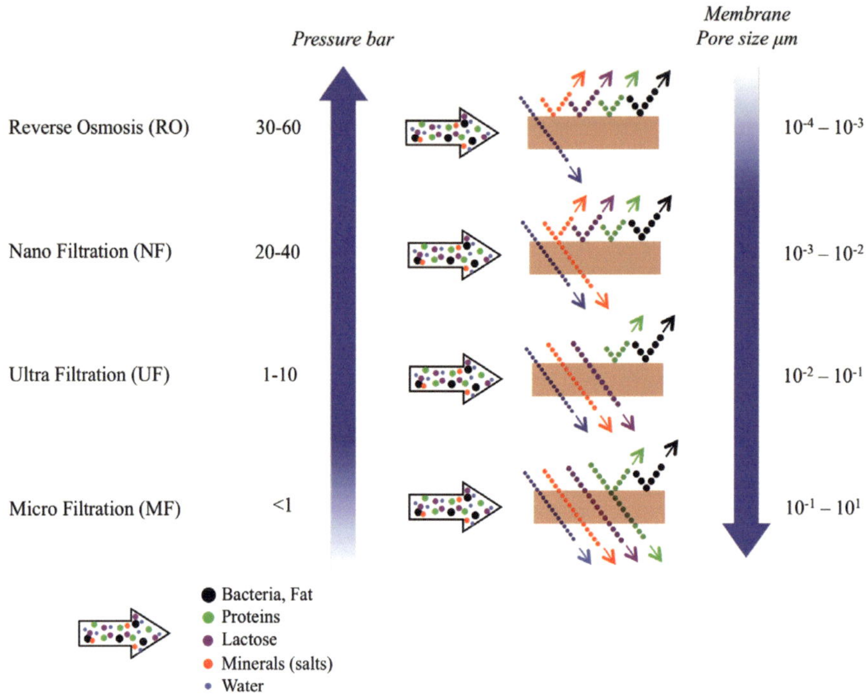

Figure 7. Principles of membrane filtration.

The energy consumption for these technological filtration (MF, UF, and NF) processes is low, as latent heat in the phase change is not consumed in the membrane separation process. Membrane separation is expected to be one of the most promising and energy-efficient separation technologies. Diffusion of solutes through non-porous biopolymer membranes is discussed using a molecular-diffusion model [75-76].

In many conventional porous membranes, the membrane material is not an active participant; only its pore structure matters, not its chemical structure [77]. A common feature of biopolymer membranes in the solution-diffusion process is that the solute molecules dissolved in the biopolymer membranes diffuse through the polymer chains (also called mass-transfer channels) and then exit the membrane at the other side phase [78]. The biopolymer is an active participant in both the solution and diffusion processes.

6.1. Diffusion in biopolymer membrane

According to a solution-diffusion mechanism based on Fick's law (Eq. 4) [79], mass transfer flux was indicated as followed:

$$J_i = -D\frac{dc_i}{dx} \tag{4}$$

http://dx.doi.org/10.5772/50697

where

J_i is the flux of component i (mol/(m^{2}s)),

D is the diffusion coefficient (m^2/s), and

dc_i/dx is the concentration gradient for component i over the length x (mol/(m^{3}m)).

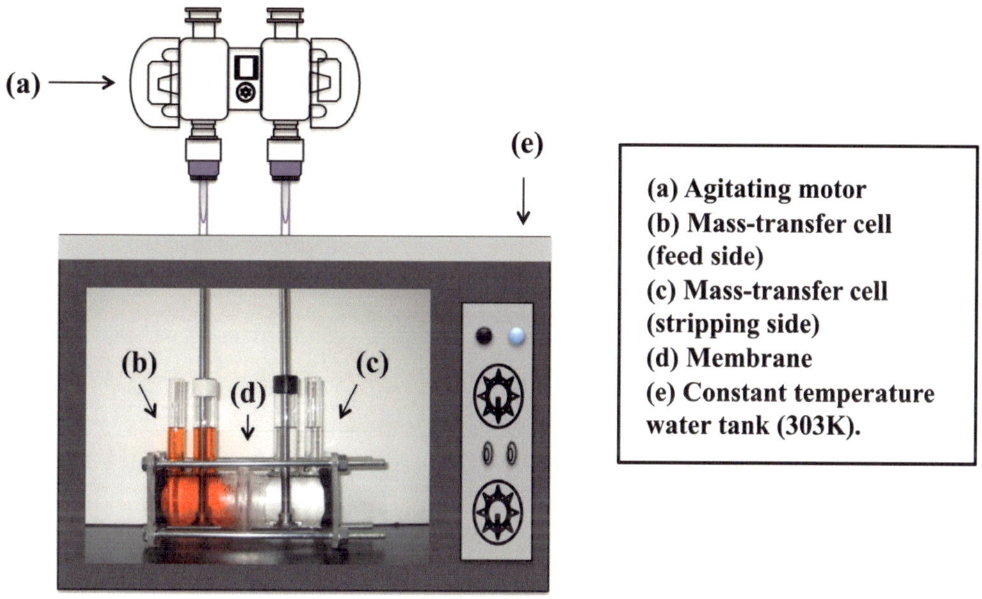

Figure 8. Schematic diagram of the mass-transfer setup in our experiment.

In this chapter, mass-transfer experiments were carried out using a standard side-by-side diffusion cell with two compartments separated by a membrane with an area of 23cm^2 (Fig. 8).

The diffusion cell was installed in a water bath to keep the temperature constant (303K). The feed compartment was filled with water-soluble marker components in solution (190ml) (Fig. 9), and the stripping compartment was filled with distilled water. During the experiment, the two compartments of solutions were stirred at a constant speed (850min^{-1}) in order to minimize the film mass-transfer resistance near the membrane surface. The solutions in the feed and stripping compartments were sampled at a fixed time interval, and the concentration was determined by measuring UV absorbance. The wavelengths of maximum absorbance are listed in Table 2. The relationship between concentration and absorbance was calibrated by taking spectra of known concentrations. The diffusion of solutes through the membrane was monitored by periodically removing 1cm^3 samples from both diffusion cells.

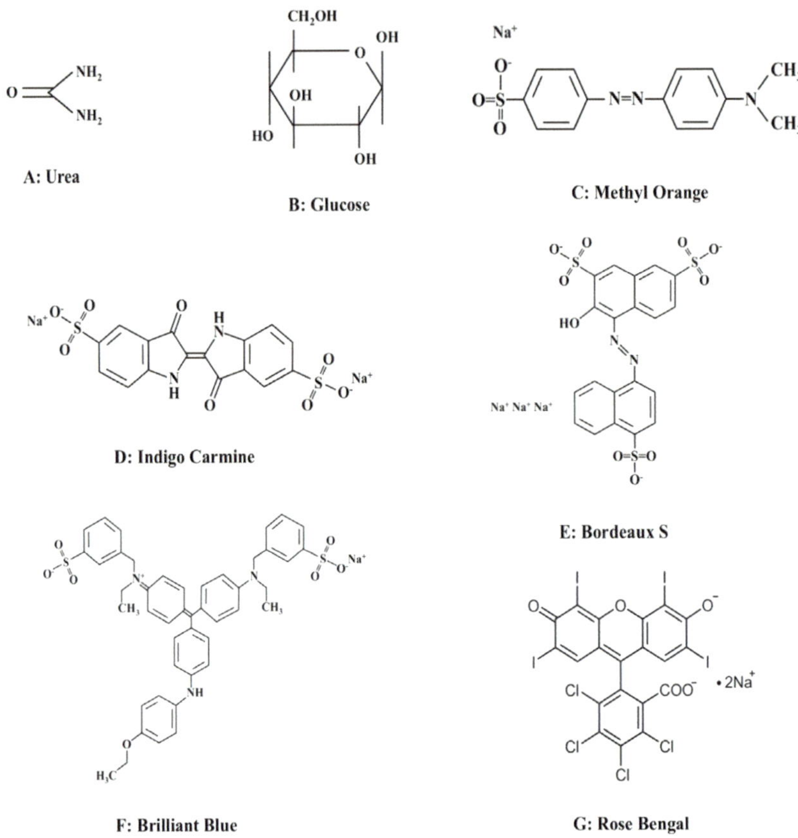

Figure 9. Chemical structure of the water-soluble components.

Marker components	Molecular Weight [Da]	Molercular Size [Å]	pH[a]	Structural Formula
Urea	60	6.0	5.4	NH_2CONH_2
Glucose	180	8.9	5.8	$C_6H_{12}O_6$
Methyl Orange	327	10.6	5.6	$C_{14}H_{14}N_9O_9SNa$
Indigo Carmine	466	11.9	5.4	$C_{16}H_8N_2Na_2O_9S_2$
Bordeaux S	604	13.0	5.9	$C_{20}H_{11}N_2Na_3O_{10}S_3$
Brilliant Blue	826	14.4	5.5	$C_{45}H_{44}N_3NaO_7S_2$
Rose Bengal	1017	15.6	5.8	$C_{20}H_2Cl_4I_4Na_2O_5$

[a] Determined from the marker component aqueous solutions of concentration at 1mM.

Table 2. The water-soluble components and their molecular size.

6.2. Determination of effective diffusion coefficient (*Deff*) in the biopolymer

The concentration of the solution transported through the membrane is required to estimate the mass-transfer characteristics of the membrane. Diffusion is a fundamental phenomenon in several physical and chemical molecular processes, representing the molecular motion of neutral or charged species in solutions [80]. The diffusion coefficient in liquid is an important parameter for understanding the complex processes of mass transfer. Several empirical methods for estimating the diffusion coefficient in aqueous phase consider infinite dilution and are based on molecular-size indicators. Figure 10 presents a schematic of the mass-transfer model. This chapter introduces the following method [81].

$$\text{Wilke \& Chang, } D_w = \frac{1.86\times 10^{-18}(\phi_B M_w)^{0.5}}{\mu_w v_A^{0.6}} \quad (5)$$

Here, D_w is the diffusion coefficient of the solute in water[$m^2\ s^{-1}$], $\mu_w$is the viscosity of water [Pa s], and φ_B is the association factor for solvent B at the required temperature T [K] (for water, φ_B=2.6). M_W is the molar mass of water [g mol^{-1}], and v_A is the molar volume of solute A at the normal boiling point [$m^3\ mol^{-1}$].

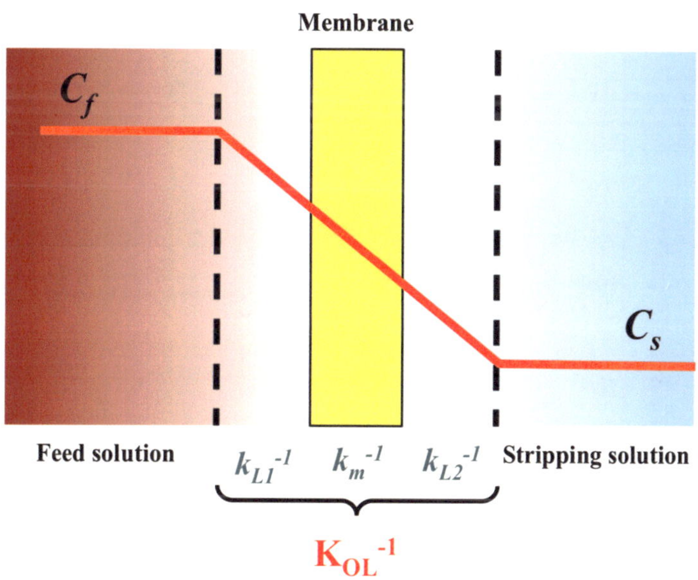

Figure 10. The schematic mass transfer model.

The concentrations in the two diffusion cells were uniform, so the mass-transfer flux was so small that the diffusion process can be regarded as in the quasi-steady state. Accordingly, we can use Eqs. (6) and (7) to calculate the effective diffusion coefficients.

$$\ln\left(1 - \frac{2C_S}{C_f}\right) = -2\frac{A}{V}K_{OL}\, t \tag{6}$$

$$K_{OL}^{-1} = k_{L1}^{-1} + k_m^{-1} + k_{L2}^{-1} \tag{7}$$

The mass-transfer resistances k_{L1}^{-1} and k_{L2}^{-1} in the overall mass transfer resistance K_{OL}^{-1} can be neglected because of the sufficiently turbulent conditions in the two diffusion cells during the experiment. K_{OL}^{-1} did not depend on the stirring rate, therefore it directly indicates the membrane mass-transfer coefficient ($k_m = Deff\, l^{-1}$). The effective diffusion coefficient in the membrane (*Deff*) was evaluated from k_m. The initial membrane thickness l in the swollen state was measured with a micrometer.

The mass-transfer characteristics were evaluated from the effective diffusion coefficient estimated by measuring the mass-transfer rate in the composite membrane. Water-soluble components were employed to determine the size of the transfer channel in the membrane. The reference molecular size was from 60 to 1017Da indicating Urea, Glucose, Methyl Orange, Indigo Carmine, Bordeaux S, Brilliant Blue, and Rose Bengal (Table 2). The diffusion coefficient (D_W) in the bulk aqueous phase was estimated by Wilke & Chang's correlation (Eq. (5)). The effective diffusion coefficient in the membrane (D_{eff}) was lower than D_W due to diffusion channels in the composite membranes (Fig. 11).

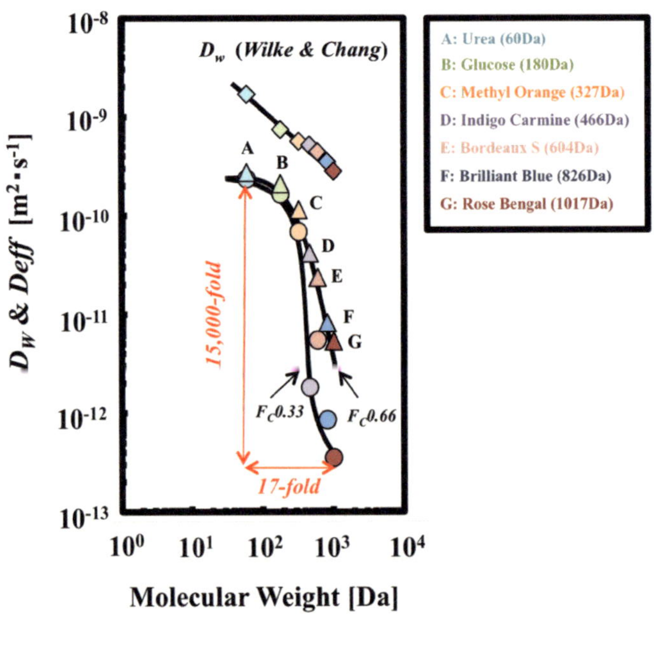

Figure 11. Effect of molecular weight on the effective diffusion coefficient of a κ-carrageenan/pullulan composite membrane.

http://dx.doi.org/10.5772/50697

The effective diffusion coefficient in the membrane (D_{eff}) changed dramatically by 15,000-fold in molecular weight when molecular weight only changed by 17-fold. The effective diffusion coefficients of the components of lower molecular weight strongly depended on the κ-carrageenan fraction F_C. D_{eff} evidently decayed under lower F_C conditions. The large dependence of D_{eff} on F_C suggests that the polymer framework becomes denser with lower F_C. In addition, there was a steep change of the effective diffusion coefficient between Methyl Orange and Indigo Carmine in each type of composite membrane. The authors therefore speculated that the mass-transfer channel was monodisperse and almost equivalent to the molecular size (11Å) of Methyl Orange.

7. Water permeability

In pressure-driven membrane separation processes such as RO and NF, solvent permeability estimation has to consider the series of resistances to fluid flux, including the membrane resistance and the boundary layer proposed, to explain the mass transfer and the hydrodynamic permeability in these processes [82]. The mass transfer inside the membrane in the absence of any osmotic effect using pure solvent (pure water) as feed indicated the moisture sensitivity of polymers. Permeability should be a more reliable indicator [83-84]. Enhancement of water permeability of the filtration membrane reduces the cost of modules used.

Goldstick [85] argued that water permeation flux in membranes follows Darcy's law for hydrodynamic flow through porous media but with swelling-pressure gradients driving the transport. In 1856, Darcy observed that the rate of flow of water through a bed of given thickness could be related to the driving pressure ΔP by the simple expression.

$$\frac{1}{A} \bullet \frac{dV}{dt} = J = \frac{\Delta P}{\eta R_m} \tag{8}$$

where J is the volumetric flux (of volume V permeating in time t through cross-section area A, m^3/m^2s) for the pressure gradient (ΔP, Pa) and the viscosity of the fluid (η, Pas); R_m refers to the permeability of the clean porous media. The resistance model is based on Darcy's law, which states that water flux through a membrane is proportional to the pressure gradient across the medium and the permeability of the medium.

If there is no fouling (clean membrane), if feed water is completely free of any solutes, and assuming laminar flow through capillary tubes of radius r, the Hagen–Poiseuille law was obtained.

$$J = \frac{\varepsilon r^2}{8\eta\tau} \frac{\Delta P}{\Delta x} \tag{9}$$

where

ε = *void fraction of the membrane (void was assumed to be cylindrical pores)* ($n\pi r^{2}$ */surface area*)

n = *number of pores*

r = *pore radius [m]*

η = *viscosity [Pa s]*

τ =*tortuosity factor*

ΔP = *trans-membrane pressure [Pa]*

Δx = *membrane thicknee [m]*

Flux is proportional to porosity, pore size, and trans-membrane pressure.

To study the performance of prepared membranes, pure-water permeability through a κ-carrageenan/pullulan membrane was measured under steady-state conditions. Prior to the experiments, the membranes were immersed in pure water for 12h and then cut into the desired size needed for fixing in a pure water permeability set-up.

The pure-water permeability experiment used a filtration cell with a volume of 200mL and effective filtration area of 21cm^2. A magnetic stirring bar was installed on the membrane upper surfaces. The filtration cell was employed for constant-flux, constant-pressure filtration. For operation in the constant-flux mode, a nitrogenous gas pump was connected to the inlet of the filtration cell and pumped the permeation water from the outlet. A pressure transducer was installed between the filtration cell and the pump in order to monitor the variation in applied pressure during filtration. The weight of the filtration water was logged by an electronic balance. The schematic of the module and set-up is presented in Fig. 12. In this chapter, the pure-water permeability was measured at different pressures and using Eq. 10.

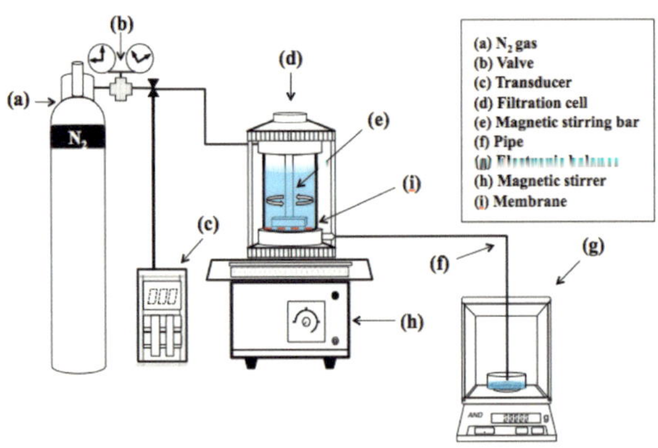

Figure 12. Schematic diagram of filtration cell used to measure steady pure-water permeability through the membranes.

$$J_V = \frac{V_P}{A\Delta t} \tag{10}$$

Here, J_V is the water flux [$m^3\ m^{-2}\ s^{-1}$], V_p is the volumetric amount of permeated water [m^3], A is the membrane area [m^2], and Δt is the sampling time [s].

The pure-water flux was measured as a function of applied pressure to investigate the stability and hydraulic properties of biopolymer membranes. In Fig. 13, the water flux and content increased linearly with increasing F_C. The result agreed with the general trend of water permeation in a hydrophilic membrane: higher water content induced higher water flux.

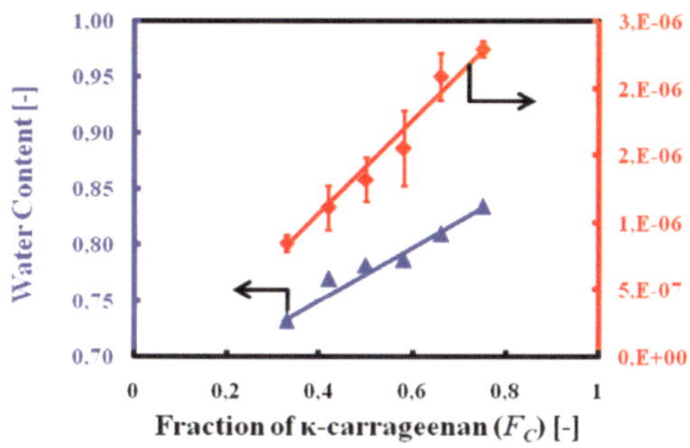

Figure 13. Change of the water flux (150KPa, 298K) and the water content of the membrane with regulated F_C values.

7.1. Obtaining the selectivity curve and molecular weight cutoff

The selectivity of a membrane is usually represented by its molecular weight cutoff [86], defined as the minimum molar mass of a test solute that is 90% retained (or 95% depending on the manufacturer) by the membrane. It is thus determined experimentally from a plot of the variation of the retention rate for tracer molecules according to their molar mass (i.e., from the selectivity (or sieving) curve) (Fig. 14).

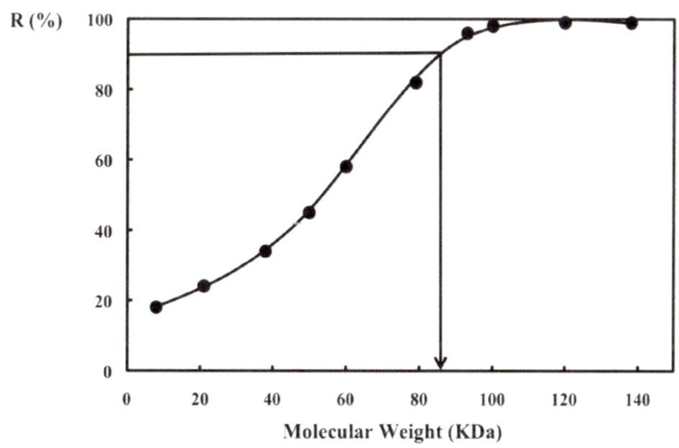

Figure 14. Example of a selectivity curve. The molecular weight cutoff, i.e., the molecular weight of a molecule rejected at 90% by the membrane is 83 kDa.

To determine a cutoff threshold, an intrinsic characteristic of the biopolymer membrane only, it is essential that the operating conditions (trans-membrane pressure, tangential circulation speed, etc.) should not affect the retention data. The rejection used for molecular weight cutoff evaluation was defined as follows.

$$R = \frac{\text{FeedConc.-PermeatConc.}}{\text{FeedConc.}} \tag{11}$$

Figure 15 presents the effect of F_C on the molecular weight cutoff of the κ-carrageenan/pullulan membranes. The molecular weight cutoff and the flux of κ-carrageenan/pullulan membranes increased with increasing F_C (Fig. 15). The molecular cutoff of F_C0.33 (F_C0.66) membrane was 327Da (466Da). The retention for high-molecular-weight tracers above 604Da was 96 to 98%.

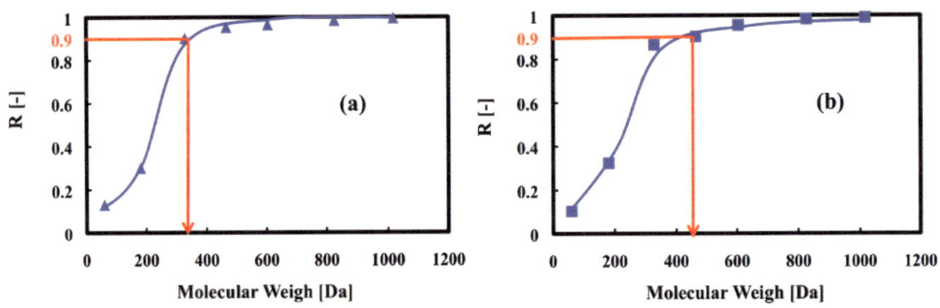

Figure 15. The selective curve of κ-carrageenan/pullulan membrane for dye molecules. (a): F_C 0.33 membrane, (b): F_C 0.66 membrane.

8. Conclusions

There are increasing reports on the physicochemical behavior of well-characterized biopolymer systems based on the fundamentals of gelation, and component interactions in the bulk and at interfaces. It appears, however, that a gap has emerged between the recent advances in fundamental knowledge and the direct application to products with a growing need for scientific input. As can be seen from the above, biopolymers are now one of the most explored potential materials for membrane-separation technology, but there is much experimental and theoretical work left to complete. An analysis of the structure–property relationships provides much information on the effects of side groups, structure, and stiffness of the main chains that can be used in directed search for advanced membrane materials of other classes. Much more interesting results have been obtained for composite and modified biopolymers. Here, significantly fewer structures have been examined, so much is yet to be done.

Biopolymer κ-carrageenan/pullulan composite membrane was successfully prepared by the casting method. It has sufficient mechanical strength for practical use and excellent mass-transfer characteristics, especially for molecular-size screening. The relationship between

http://dx.doi.org/10.5772/50697

mass-transfer characteristics and the mass fraction of κ-carrageenan in the composite membrane was formulated based on mass-transfer flux and pure-water flux experiments. The results provided a novel and simple method of preparing membranes and mass-transfer channels based on molecular-size indicators, and suggested that different F_C values significantly affect the mass-transfer permeability. The water permeation flux as a function of applied pressure provided valuable technical information for investigating the stability and hydraulic properties of the composite membranes. It was concluded that κ-carrageenan/pullulan composite membranes with a cross-linked hydrophilic structure exhibited high selectivity and high water flux. Thus, mass-transfer investigations are very useful and informative for studying and analyzing composite membranes.

Acknowledgements

This work was supported by research funding grants provided by the Iijima Memorial Foundation for the Promotion of Food Science and Technology. The authors sincerely thank Dr. Kei Tao of Nihon University, who provided technical assistance on SEM photography.

Author details

Peng Wu and Masanao Imai*

*Address all correspondence to: XLT05104@nifty.com

Course in Bioresource Utilization Sciences, Graduate School of Bioresource Sciences, Nihon University, Japan

References

[1] Krishna, R. K. S. V., Vijaya, B., Naidu, K., Subha, M. C. S., Sairam, M., Mallikarjuna, N. N., & Aminabahvi, T. M. (2006). Novel carbonhydrate polymeric blend membranes in pervaporation dehydration of acetic acid. *Carbonhydrate Polymer*, 66, 345-351.

[2] Stanek, L. G., Heilmann, S. M., & Gleason, W. B. (2006). Preparation and copolymerization of a novel carbohydrate containing monomer. *Carbonhydrate Polymer*, 65, 552-556.

[3] Lee, K. P., Arnot, T. C., & Mattia, D. (2011). A review of reverse osmosis membrane materials for desalination-Development to date and future potential. *Journal of Membrane Science*, 370, 1-22.

[4] Catherine, C. (2009). A review of membrane processes and renewable energies for desalination. *Desalination*, 245, 214-231.

[5] Guessasma, S., Hamdi, A., & Lourdin, D. (2009). Linear modeling of biopolymer systems and related mechanical properties. *Carbonhydrate Polymer*, 76, 381-388.

[6] Imeson, A. (2009). Food Stabilisers, Thickeners and Gelling Agents. Wiley-Blackwell, 343, 10.1002/9781444314724.

[7] Kumar, A., Srivastava, A., Galaev, I. Y., & Mattiasson, B. (2007). Smart polymer: Physical forms and bioengineering applications. *Prog. Polym. Sci.*, 32, 1205-1237.

[8] Clark, A. H. (1996). Biopolymer gels. *Current Opinion in Colloid & Interface Science*, 1, 712-717.

[9] McCray, S. B., Vilker, V. L., & Nobe, K. (1991). Reverse osmosis cellulose acetate membranes. I. Rate of hydrolysis. *Journal of Membrane Science*, 59, 305-316.

[10] Rinaudo, M. (2006). Chitin and chitosan: Properties and applications. *Prog. Polym. Sci.*, 31, 603-632.

[11] Pillai, C. K. S., Paul, W., & Sharma, C. P. (2009). Chitin and chitosan polymers: Chemistry, solubility and fiber formation. *Progress in polymer Science*, 34, 641-678.

[12] Lee, Y. M., & Shin, E. M. (1991). Pervaporation separation of water-ethanol through modified chitosan membranes. IV. Phosphorylated chitosan membranes. *Journal of Membrane Science*, 64, 145-152.

[13] Won, W., Feng, X. S., & Lawless, D. (2002). Pervaporation with chitosan membranes: separation of dimethyl carbonate/methanol/water mixtures. *Journal of Membrane Science*, 209, 493-508.

[14] Devi, D. A., Smitha, B., Sridhar, S., & Aminabhavi, T. M. (2005). Pervaporation separation of isopropanol/water mixtures through crosslinked chitosan membranes. *Journal of Membrane Science*, 262, 91-99.

[15] Huang, X. J., Ge, D., & Xu, Z. K. (2007). Preparation and characterization of stable chitosan nanofibrous membrane for lipase immobilization. *European Polymer Journal*, 43, 3710-3718.

[16] Beppu, M. M., Vieira, R. S., Aimoli, C. G., & Santana, C. C. (2007). Crosslinking of chitosan membranes using glutaraldehyde: Effect on ion permeability and water absorption. *Journal of Membrane Science*, 301, 126-130.

[17] Zeng, X. F., & Ruckenstein, E. (1998). Cross-linked macroporous chitosan anion-exchange membranes for protein separations. *Journal of Membrane Science*, 148, 195-205.

[18] Takahashi, T., Imai, M., Suzuki, I., & Sawai, J. (2008). Growth inhibitory effect on bacteria of chitosan membranes regulated with deacetylation degree. *Biochemical Engineering Journal*, 40, 485-491.

[19] Takahashi, T., Imai, M., & Suzuki, I. (2007). Water permeability of chitosan membrane involved in deacetylation degree control. *Biochemical Engineering Journal*, 36, 43-48.

[20] Kashima, K., Imai, M., & Suzuki, I. (2010). Superior molecular size screening and mass-transfer characterization of calcium alginate membrane. *Desalination and water treatment*, 17, 143-159.

[21] Kashima, K., & Imai, M. (2011). Dominant impact of the α-L-guluronic acid chain on regulation of the mass transfer character of calcium alginate membranes. *Desalination and water treatment*, 134, 257-265.

[22] El Gamal, A. A. (2010). Biological importance of marine algae. *Saudi Pharmaceutical Journal*, 18, 1-25.

[23] Rasmussen, R. S., & Morrissey, M. T. (2007). Marine biotechnology for production of food ingredients. *Advances in Food and Nutrition Research*, 52, 237-292.

[24] Millane, R. P., Chandrasekaran, R., & Arnott, S. (1988). The molecular structure of kappa-carrageenan and comparison with iota-carrageenan. *Carbohydrate Research*, 183, 1-17.

[25] Dea, I. C. M., Mckinnon, A. A., & Rees, D. A. (1972). Tertiary and Quaternary Structure in Aqueous Polysaccharide Systems which Model Cell Wall Cohesion: Reversible Changes in Conformation and Association of Agarose, Carrageenan and Galactomannans. *Journal of Molecular Biology*, 68, 153-172.

[26] Bixler, H. J., Johndro, K., & Falshaw, R. (2001). Kappa-2 carrageenan: structure and performance of commercial extracts II. Performance in two simulated dairy applications. *Food Hydrocolloids*, 15, 619-630.

[27] Falshaw, R., Bixler, H. J., & Johndro, K. (2001). Structure and performance of commercial kappa-2 carrageenan extracts I. Structure analysis. *Food Hydrocolloids*, 15, 441-452.

[28] Ekstrom, A. G., & Kuivinen, J. (1983). Molecular weight distribution and hydrolysis behaviour of carrageenans. *Carbohydrate Research*, 116, 89-94.

[29] El Gamal, A. A. (2010). Biological importance of marine algae. *Saudi Pharmaceutical Journal*, 18, 1-25.

[30] Mangione, M. R., Giacomazza, D., Bulone, D., Martorana, V., Cavallaro, G., & San Biagio, P. L. (2005). K^+ and Na^+ effects on the gelation properties of k-Carrageenan. *Biophysical Chemistry*, 113, 129-135.

[31] Leathers, T. D. (2003). Biotechnological production and applications of pullulan. *Appl Microbiol Biotechnol.*, 62, 468-473.

[32] Ueda, S., Fujita, K., Komatsu, K., & Nakashima, Z. (1963). Polysaccharide produced by the genus Pullularia I. Production of polysaccharide by growing cells. *Applied Microbiology*, 11, 211-215.

[33] Singh, R. S., Saini, G. K., & Kennedy, J. F. (2008). Pullulan: Microbial sources, production and applications. *Carbohydrate Polymers*, 73, 515-531.

[34] Xiao, Q., Lim, L. T., & Tong, Q. (2011). Properties of pullulan-based blend film as affected by alginate content and relative humidity. *Carbohydrate Polymers*, 87, 227-234.

[35] Shih, F. F., Daigle, K. W., & Champagne, E. T. (2011). Effect of rice wax on water vapour permeability and sorption properties of edible pullulan films. *Food Chemistry*, 127, 118-121.

[36] Lazaridou, A., Biliaderis, C. G., & Kontogiorgos, V. (2003). Molecular weight effects on solution rheology of pullulan and mechanical properties of its films. *Carbohydrate Polymers*, 52, 151-166.

[37] Trinetta, V., Cutter, C. N., & Floros, J. D. (2011). Effects of ingredient composite on optical and mechanical properties of pullulan film for food-packaging applications. *LWT- Food Science and Technology*, 44, 2296-2301.

[38] Ulbricht, M. (2006). Advanced functional polymer membranes. *Polymer*, 47, 2217-2262.

[39] Lebrun, L., Blanco, J. F., & Metayer, M. (2005). Preparation of ion-exchange membranes using pullulan as polymer matrix. *Carbohydrate Polymers*, 61, 1-4.

[40] Datta, S., Mody, K., Gopalsamy, G., & Jha, B. (2011). Novel application of κ-carrageenan: As a gelling agent in microbiological media to study biodiversity of extreme alkaliphiles. 85, 465-468.

[41] Wu, P., & Imai, M. (2011). Food polymer pullulan-κ-carrageenan composite membrane performed smart function both on mass transfer and molecular size recognition. *Desalination and Water Treatment*, 34, 239-245.

[42] Sairam, M., Sereewatthanawut, E., Li, K., Bismarck, A., & Livingston, A. G. (2011). Method for the preparation of cellulose acetate flat sheet composite membranes for forward osmosis-Desalination using $MgSO_4$ draw solution. *Desalination*, 273, 299-307.

[43] Miao, J., Chen, G., Gao, C., & Dong, S. X. (2008). Preparation and characterization of N,O-carboxymethyl chitosan/Polysulfone composite nanofiltration membrane crosslinked with epichlorohydrin. *Desalination*, 233, 147-156.

[44] Peeva, P. D., Million, N., & Ulbricht, M. (2012). Factors affecting the sieving behavior of anti-fouling thin-layer cross-linked hydrogel polyethersulfone composite ultrafiltration membranes. *Journal of Membrane Science*, 390-391, 99-112.

[45] Elimelech, M., Zhu, X. H., Childress, A. E., & Hong, C. S. (1997). Role of membrane surface morphology in colloidal fouling of cellulose acetate and composite aromatic polyamide reverse osmosis membranes. *Journal of Membrane Science*, 127, 101-109.

[46] Zio, A. D., Prisciandaro, M., & Barda, D. (2005). Disinfection of surface waters with UF membranes. *Desalination*, 179, 297-305.

[47] Singh, G., Rana, D., Matsuura, T., Ramakrishna, S., Narbaitz, R. M., & Tabe, S. (2010). Removal of disinfection byproducts from water by carbonized electrospun nanofibrous membranes. *Separation and Purification Technology*, 74, 202-212.

[48] Lameloise, M., Matinier, H., & Fargues, C. (2009). Concentration and purification of malate ion from a beverage industry waste water using electrodialysis with homopolar membranes. *Journal of Membrane Science*, 343, 73-81.

[49] Yushina, Y., & Hasegawa, J. (1994). Process performance comparison of membrane introduced anaerobic digestion using food industry waste water. *Desalination*, 98, 413-421.

[50] Blocher, C., Noronha, M., Funfrocken, L., Dorda, J., Mavrov, V., Janke, H. D., & Chmiel, H. (2002). Recycling of spent process water in the food industry by an integrated process of biological treatment and membrane separation. *Desalination*, 144, 143-150.

[51] Alp, B., Mutltu, S., & Mutlu, M. (2000). Glow-discharge-treated cellulose acetate (CA) membrane for a high linearity single-layer glucose electrode in the food industry. *Food Research International*, 33, 107-112.

[52] Kabay, N., Bryjak, M., Schlosser, S., Kitis, M., Avlonitis, S., Matejka, Z., Al-Mutaz, I., & Yuksel, M. (2008). Adsorption-membrane filtration (AMF) hybrid process for boron removal from seawater: an overview. *Desalination*, 223, 38-48.

[53] Redondo, J., Busch, M., & Witte, J. D. (2003). Boron removal from seawater using FILMTECTM high rejection SWRO membranes. *Desalination*, 156, 229-238.

[54] Hu, J. H., Song, L. F., Ong, S. L., Phua, E. T., & Ng, W. J. (2005). Biofiltration pretreatment for reverse osmosis (RO) membrane in a water reclamation system. *Chemosphere*, 59, 127-133.

[55] Alvarez-Hornos, F. J., Volckaert, D., Heynderickx, P. M., & Langenhove, H. V. (2011). Performance of a composite membrane bioreactor for the removal of ethyl acetate from waste air. *Bioresource Technology*, 102, 8893-8898.

[56] Zheng, X., & Liu, J. (2006). Dyeing and printing wastewater treatment using a membrane bioreactor with a gravity drain. *Desalination*, 190, 277-286.

[57] Su, M., Teoh, M. M., Wang, K. Y., Su, J. C., & Chung, T. S. (2010). Effect of inner-layer thermal conductivity on flux enhancement of dual-layer hollow fiber membranes in direct contact membrane distillation. *Journal of Membrane Science*, 364, 278-289.

[58] Shah, B. G., Shahi, V. K., Thampy, S. K., Rangarajan, R., & Ghosh, P. K. (2009). Comparative studies on performance of interpolymer and heterogeneous ion-exchange membranes for water desalination by electrodialysis. *Desalination*, 172, 257-265.

[59] Misdan, N., Lau, W. J., & Ismail, A. F. (2012). Seawater Reverse Osmosis (SWRO) desalination by thin-film composite membrane-Current development, challenges and future prospects. *Desalination*, 287, 228-237.

[60] Maguire-Boyle, S. J., & Barron, A. R. (2011). A new functionalization strategy for oil/water separation membranes. *Journal of Membrane Science*, 382, 107-115.

[61] Steward, P. A., Hearn, J., & Wilkinson, M. C. (2000). An overview of polymer latex film formation and properties. *Advances in Colloid and Interface Science*, 86, 195-267.

[62] Tong, Q., Xiao, Q., & Lim, L. T. (2008). Preparation and properties of pullulan-alginate-carboxymethylcellulose blend films. *Food Research International*, 41, 1007-1014.

[63] Martelli, S. M., Moore, G. R. P., & Laurindo, J. B. (2006). Mechanical Properties, Water Vapor Permeability and Water Affinity of Feather Keratin Films Plasticized with Sorbitol. *J. Polym. Environ.*, 14, 215-222.

[64] Rao, P. S., Smitha, B. S., Sridhar, S., & Krishnaiah, A. (2006). Preparation and performance of poly(vinyl alcohol)/polyethyleneimine blend membranes for the dehydration of 1,4-dioxane by pervaporation: Comparison with glutaraldehyde cross-linked membranes. *Separation and Purification Technology*, 48, 244-254.

[65] Takahashi, T., Imai, M., & Suzuki, I. (2008). Cellular structure in an N-acetyl-chitosan membrane regulate water permeability. *Biochemical Engineering Journal*, 42, 20-27.

[66] McCray, S. B., Vilker, V. L., & Nobe, K. (1991). Reverse osmosis cellulose acetate membranes. I. Rate of hydrolysis. *Journal of Membrane Science*, 59, 305-316.

[67] Vasileva, N., & Godjevargova, T. (2004). Study on the behaviour of glucose oxidase immobilized on microfiltration polyamide membrane. *Journal of Membrane Science*, 239, 157-161.

[68] Villegas, M., Vidaurre, E. F., Habert, A. C., & Gottifredi, J. C. (2011). Sorption and pervaporation with poly(3-hydroxybutyrate) membranes: methanol/methyl tert-butyl ether mixtures. *Journal of Membrane Science*, 367, 103-109.

[69] Matsuoka, Y., Kanda, N., Lee, Y. M., & Higuchi, A. (2006). Chiral separation of phenylalanine in ultrafiltration through DNA-immobilized chitosan membranes. *Journal of Membrane Science*, 280, 116-123.

[70] Papageorgiou, S. K., Katsaros, F. K., Favvas, E. P., Romanos, G. E., Athanasekou, C. P., Beltsios, K. G., Tzialla, O. I., & Falaras, P. (2012). Alginate fibers as photocatalyst immobilizing agents applied in hybrid photocatalytic/ultrafiltration water treatment processes. *Water Research*, 46, 1858-1872.

[71] Miao, J., Chen, G. H., & Gao, C. J. (2005). A novel kind of amphoteric composite nanofiltration membrane prepared from sulfated chitosan (SCS). *Desalination*, 181, 173-183.

[72] Li, X. L., Zhu, L. P., Zhu, B. K., & Xu, Y. Y. (2011). High-flux and anti-fouling cellulose nanofiltration membranes prepared via phase inversion with ionic liquid as solvent. *Separation and Purification Technology*, 83, 66-73.

[73] Wu, J., & Yuan, Q. (2002). Gas permeability of a novel cellulose membrane. *Journal of Membrane Science*, 204, 185-194.

[74] Xiao, S., Feng, X. S., & Huang, R. Y. M. (2007). Trimesoyl chloride crosslinked chitosan membranes for CO2/N2 separation and pervaporation dehydration of isopropanol. *Journal of Membrane Science*, 306, 36-46.

[75] Feil, H., Bae, Y. H., Feijen, J., & Kim, S. W. (1991). Molecular separation by thermosensitive hydrogel membranes. *Journal of Membrane Science*, 64, 283-294.

[76] Peppas, N. A., & Reinhart, C. (1983). Solute Diffusion in Swollen Membranes. Part I. A New Theory. *Journal of Membrane Science*, 15, 275-287.

[77] Baltus, R. E. (1997). Characterization of the pore area distribution in porous membranes using transport measurements. *Journal of Membrane Science*, 123, 165-184.

[78] Krajewska, B., & Olech, A. (1996). Pore structure of gel chitosan membranes. I. Solute diffusion measurements. *Polymer Gels and Networks*, 4, 33-43.

[79] Neogi, P. (1996). Diffusion in Polymer. Marcel Dekker, 309.

[80] Chen, C. X., Han, B. B., Li, J. D., Shang, T. G., & Jiang, W. J. (2001). A new model on the diffusion of small molecule penetrants in dense polymer membranes. *Journal of Membrane Science*, 187, 109-118.

[81] Miyabe, K., & Isogai, R. (2011). Estimation of molecular diffusivity in liquid phase systems by the Wilke-Chang equation. *Journal of Chromatography A*, 1218, 6639-6645.

[82] Mehdizadeh, H., Molaiee-Nejad, K., & Chong, Y. C. (2005). Modeling of mass transport of aqueous solutions of multi-solute organics through reverse osmosis membranes in case of solute-membrane affinity Part 1. Model development and simulation. *Journal of Membrane Science*, 267, 27-40.

[83] Yaroshchuk, A. E. (1995). Solution-diffusion-imperfection model revised. *Journal of Membrane Science*, 101, 83-87.

[84] Yaroshchuk, A. E. (1995). The role of imperfections in the solute transfer in nanofiltration. *Journal of Membrane Science*, 239, 9-15.

[85] Fatt, I., & Goldstick, T. K. (1965). Dynamics of water transport in swelling membranes. *J. Colloid Sci.*, 20, 962-989.

[86] Jonsson, G. (1985). Molecular weight cut-off for ultrafiltration membranes of varing pore size. *Desalination*, 53, 3-10.

Section 2

RO Process Chemistry and Control

Chapter 4

Chemistry in the Operation and Maintenance of Reverse Osmosis Systems

Ph.D. Robert Y. Ning

Additional information is available at the end of the chapter

http://dx.doi.org/10.5772/39385

1. Introduction

Across the spectrum of industrial and municipal water utilization and treatment plants, extensive desalination and purification of water relies on the use of reverse osmosis (RO) membranes. Sustaining the productivity of RO plants as continuous processes for water purification has been since the late 1970s, and still is, a significant technological challenge. The challenge is magnified on the one hand by the increasing shortages of water thus driving down the quality of available raw waters, and on the other hand by the demand and the high cost of lost production that can result from insufficient productivity of RO systems. Reduced productivity of RO plants exerts serious economic impact on the downstream production of steam, power, microelectronics, pharmaceuticals and beverages among other products. Not only used in the front-end to provide supply of high quality process water, the loss of RO capacity to process wastewater at the back-end to allow regulated discharges can shut down production or operation of some industrial complexes. With all these requirements, efficient operation and maintenance (O&M) of RO plants based on an understanding of chemistry is essential.

The design and working of an RO system as a unit operation is widely described in articles, books, technical literature and design software of membrane manufacturers and updated versions of user association manuals, such as from the American Water Works Association [1] The sensitivity of RO membranes towards fouling however has presented great challenges and crises to O&M personnel [2]. Once a plant is built, changes in source water quality, inadequacies in pretreatment unit operations, and inappropriate O&M procedures lead to costly repairs to the RO. Even more serious, is the stoppage of water supply to the entire production plant of high value products. Insufficient attention to changing chemistry in raw water, and inadequate performance of pretreatment units result in our current industry practice of expecting 2 to 4 years membrane service life along with frequent stoppages for membrane cleaning and system maintenance. When optimally controlled, RO membranes have lasted more than 12 years. Systems exist that have operated continuously, and not have to be cleaned for many years. In this article, we point to aspects of process chemistry peculiar to RO plants [3-21].

2. Membrane Fouling Mechanisms

Feedwaters to RO systems typically are concentrated by a factor of 2 - 10 (50%-90% recovery) during production of permeate water. For simplicity in classification, three classes of fouling can be said to occur [8-11].

2.1. Firstly, there is scaling

The solubility limits of various dissolved salts in the concentrate stream may be exceeded, leading to deposition and growth of crystals in the flow channels and membrane surface of the RO elements. This type of fouling is referred to as scaling. Most common examples of scales are calcium carbonate, sulfates of calcium, strontium and barium, and calcium fluoride and calcium phosphate. In the examination of foulants by naked eye or with magnifying glass, crystals have well defined shapes. Inhibitors injected continuously into feedwaters to suppress crystallization are called antiscalants. For scaling to occur, seed crystals from in the super-saturated concentrate. The seed crystals may grow into discernable shapes such as plates, flakes, prisms or needles, or remain as finely dispersed particles, visible or invisible to the naked eyes. Antiscalants work by inhibiting the growth of such seed crystals, and forcing the RO concentrate to remain for a time in a supersaturated state. This mechanism is known as threshold inhibition. This task of scale growth inhibition is efficiently accomplished by low concentrations of antiscalants maintaining high super-concentrations in the RO concentrate. The capabilities of certain antiscalants in controlling the most common scales are listed in Table 1. Such antiscalants can be used to effectively replace traditional pretreatment methods of removing the offending foulants from the raw water stream [4-7, 21].

2.2. Secondly, there is colloidal fouling

The foulants appear typically as colorless to yellow or brown soft amorphous layer during autopsy when membranes are cut and unrolled for visual examination [10]. Raw waters contain heavy loads of particles ranging from the visible kind to the smallest of the invisible. Colloidal particles can be considered here as less than 1.0 micron in size. Below 1 micron, they are invisible to the naked eye, nearly undetected by turbidity and Silt Density Index (SDI) measurements. The 0.45 micron filters used in SDI measurement do not retain colloidal particles. In fact even microfiltration and ultrafiltration pretreatment in RO systems still pass colloidal particles that result in severe colloidal fouling of the RO membranes [14,18,19]. Elemental composition analyses of washed and dried colloidal foulants for elemental carbon, hydrogen and nitrogen generally show predominance of complex microbial matter from natural sources. Inorganic components clearly discernable are clays (aluminum silicate), silt (ferric-aluminium-magnesium silicates), silica (polymer represented by composition of SiO2), and ferric and aluminum oxyhydroxides when such are used as coagulants for water clarification upstream [10, and unpublished data].

RO Foulant	Traditional Pretreatment	Current Capability
1. Calcium Carbonate	Acidification to lower LSI, then with antiscalant to maximum LSI=2.5	Antiscalant alone: LSI=3.3;S&DSI="/4.5
2. Iron, manganese	Oxidation/filtration: Greensand, manganese dioxide, catalytic oxid'n	Antiscalant alone: Fe and Mn at "/ 8ppm
3. Silica: Reactive	Lime, Ion-Exchange (OH)	Antiscalant: "/280ppm
Non-reactive	None	Antifoulant: variable
4. Calcium Sulfate	Lime, Ion-Exchange	Antiscalant: "/400x sat'n
5. Strontium Sulfate	Ion-Exch ange	Antiscalant: "/43x sat'n
6. Barium Sulfate	Ion-Exchange	Antiscalant: "/51x sat'n
7. Calcium Fluoride	Lime, Ion-Exchange	Antiscalant: "/16,000x sat'n
8. Colloidal Organic Matter and Sulfur	UF, MF, coagulation/clarification/MMF, slow sand and carbon filter	Antifoulant alone

Table 1. RO System Design Advances Made Possible by New Antiscalant and Antifoulant Chemicals.

Colloidal matter pre-existing in feedwater can aggregate and adhere to the membrane and brine flow channels due to increased concentration, salinity, compaction, flocculation, surface interactions and other physical and chemical factors [11, 17-19]. The colloids can become organic or inorganic or composite flocs. Dissolved ionic ferric, aluminum and calcium hydroxides and silicic acid grow to increasing particle sizes by polymerization [3,17,18]. Cross-linking, and complexation of organic and inorganic polymers become gels and amorphous foulants commonly seen on membranes [10,11]. Biotic debris such as polysaccharides and dead cellular matter contribute largely to this type of foulants. Through solving numerous fouling problems in existing RO plants, it has become obvious that excessive and inappropriate application of pretreatment chemicals aggravate the tendency of natural colloids present in RO feedwaters to coagulate and become foulants. Anticoagulants and anti-deposition agents recently developed show promise in inhibiting this fouling process [14]. The term antifoulants then can be a term applied specifically in this particulate fouling context.

2.3. Thirdly

Biofouling is a prominent source of fouling. True of all water treatment or distribution systems, is the growth and anchoring of microorganisms. Moderate temperatures and minimal nutrient levels in RO raw waters can support at times explosive growths of microorganisms. Bacteria capable of cell division every 20 minutes can grow from a normal count per unit volume of water to millions in the period of an 8 hours shift. Due to the tendency of bacteria to secrete polymers that anchor themselves to surfaces to facilitate growth as the biofilm, this fouling mechanism is unique and poses a serious threat to the operation of RO systems. This threat is compounded by the great difficulty of treating and completely removing biofilm from the membrane surface. Associated with biofouling of proliferating nature, is the large presence of non-proliferating organic materials secreted or deposited in natural waters. Microbes and planktons secrete into their water environment large amounts of exocellular polysaccharides in the sunlit surface waters on land and at sea, then remain in the waters as cellular debris. For RO systems maintenance, biofouling then has the two separate component of biofilm proliferation and the colloidal organic fouling by mobile colloidal particles.

2.4. An over-riding effect

Involving coagulation and deposition of colloidal particles in the RO system parallels the delta effects of large river systems. Finer and finer particles in river water that survive physical deposition during the river's meandering course reaches the sea. When the low salinity (low Total Dissolved Solids-TDS) river water meets the high TDS seawater, colloidal particles coagulate and precipitate to form the river delta. This is a model of what happens in the last stages of the RO membrane system where concentrations of colloidal particles and TDS rise rapidly, and colloidal fouling takes place.

3. Antifoulant Chemical Design and Application

The term antifoulant used here is in its broadest meaning covering scaling, particulate fouling and microbial fouling- the three classes of fouling mechanisms discussed above. Strategies aimed at controlling each type of fouling is summarized here.

3.1. For scale control

The development and application of antiscalants is well known and reviewed in the field of boiling water and cooling water chemistry, and applied to boilers, evaporators, cooling towers and cooling systems. Anionic polymers, polyphosphates and organo-phosphorous compounds, sometimes referred to as threshold inhibitors and dispersants, are used in sub-stoichiometric amounts, usually in the range of 1-5 mg/liter concentrations in RO systems. By binding to surfaces of growing crystal nuclei, the rates of crystallization from supersaturated solutions are retarded, and crystal-packing orders are modified. By this mechanism, crystallization rates are so retarded that although super-saturation of solutes in the water will eventually equilibrate through crystallization, within the residence time of the water in the system, there is little or no scale formation. The uniqueness of RO among water conditioning systems is that the residence time is very short (a few seconds), concentration of seed crystals is low, and temperature is constant. For this reason, higher levels of super-saturation without crystallization are possible. On the other hand, the limits of saturation and rates of scaling are hard to model, measure and predict. Interferences come from other solutes in the water, organic or inorganic. Assumptions of RO fouling limits vary considerably among practitioners.

http://dx.doi.org/10.5772/39385

3.2. For controlling colloidal fouling

The task is much more challenging due to the variety of types of potential foulants and the complexity of their interactions [3,4,8-11,13,14, 17-19], with each other in the same water, and with the membrane. Stability and agglomeration of colloidal particles is a subject of major importance in natural waters as well as in the treatment of process waters [14]. Drawing on the basic science of colloids, and testing of model foulants suggested by RO foulant analysis data [10,14,19], progress is made steadily with the development of antifoulants.

3.3. Concerning prevention and management of bio-fouling

In water treatment systems, the literature is extensive. Much of the art and science found useful, is applicable to RO systems as well. Several factors peculiar to the RO system can be mentioned. Chemicals used to sanitize and clean the system have to be chemically compatible with the thin, salt-rejecting, polyamide or cellulose acetate barrier membrane. Of prime concern is that accumulation and exponential growth of the microorganisms should not be allowed to occur within the system. Pretreatment of feedwater, adequate maintenance of upstream unit operations, continuous flow of water through the RO unit, good monitoring and sanitization program, and used of preservatives during downtime [15] are important to this end. Normalized permeate flow and differential pressure in the system [20] are sensitive indicators of bio-fouling.

4. Membrane Cleaning

Practical procedure for maintenance cleaning is limited to the re-circulation of cleaning solutions through the membrane elements. By a patented method [16] of membrane reconditioning, spiral wound elements with the hard casing removed, are routinely used in selecting effective cleaners for cleaning by re-circulation. This allows for visual inspection of membrane surface after each cleaning test. It is apparent that except for easily soluble foulants like calcium carbonate, calcium phosphate or ferric and aluminum hydroxide, and less easily dissolved foulants like silica, silicates, calcium sulfate and strontium sulfate, nearly all other RO foulants are only slightly dissolved in even the best matched cleaning solutions with extensive soaking. Removal requires high tangential flow velocities to facilitate detachment of foulants from the surfaces of the membrane and the brine-side spacer screen. Cleaning is usually partially effective, especially where flow channels are clogged, and large patches within the elements are inaccessible to the re-circulating cleaning solution. For this reason, the need for cleaning should be minimized or completely eliminated by the new antiscalants and antifoulants now available, and adequate pretreatment and pilot testing of cleaning process developed during pilot testing stage. When cleaning is necessary during operation, it should be performed at the earliest stages of fouling.

It is generally agreed among membrane manufacturers and practitioners that RO systems should be cleaned before the following performance changes are reached:

1. Loss of 10 to 15% in normalized permeate flowrate.

2. Increase of 10 to 15% in differential pressure.

3. Decrease of 1 to 2% in salt rejection.

If a cleaning procedure fails to fully restore the system performance to the reference RO system startup values, it is certain that continued use of the same cleaning procedure will lead to accelerating decline in system performance and increasing cleaning frequency. For this reason, it is important to address two issues at this point: a) find an improved cleaning procedure, b) investigate possible improvement of pretreatment to avoid membrane fouling. Continue cleaning and process improvement efforts until stability of the RO performance is attained. Even with well piloted and designed RO plants, and smooth operations initially, source water qualities invariably change over time. Equipment and personnel changes also impact performance, requiring constant vigil and preparedness for continuous improvement of the plant.

4.1. Choosing Cleaners

Major membrane manufacturers generally define five types of foulants for which various generic chemicals are recommended for blending at the site where cleaning solutions are prepared. The five types of foulants are: 1) Acid-soluble Foulants, 2) Bio-film/Bacterial Slime/Biological Matter, 3) Carbon-containing Oils/Organic Matter, 4) Dual Organic and Inorganic Coagulated Colloids, and 5) Silica and Silicates. Proprietary booster cleaners are commercially available to fortify the effectiveness of these generic cleaners that are formulated at the site. For convenience and technical support, a large variety of proprietary RO membrane cleaners are available from chemical suppliers that specialize in RO operations. Such proprietary cleaners and cleaning support are available when generic cleaners do not perform adequately.

4.2. Cleaning Strategies

Experience has shown that within the same class of foulants, responses to the same cleaning solution can vary considerably. Elemental analyses of foulants and cleaning studies have shown that more than one type of foulant can be present on the membrane at the same time, requiring sequential cleaning with different cleaners. Sometimes even the order of cleaners used would make a significant difference. All this is to say that the choice of cleaners and the cleaning procedure to be used is an empirical science. For a given set of conditions in a plant, cleaning efficiencies are improved by trials over time. The progress of improvement can be greatly accelerated by conducting off-line cleaning studies on test skids for single fouled elements taken from the plant.

http://dx.doi.org/10.5772/39385

When a better cleaning method is needed in the plant, the following are the alternative strategies:

Strategy 1: The plant has a history of using generic cleaning chemicals, and modest improvement in effectiveness is needed, consider purchase of proprietary booster cleaners.

Strategy 2: A significant cleaning improvement is needed, look for proprietary cleaner supplier with associated cleaning expertise. Option 1: With prior knowledge of the characteristics of the foulant on hand, with consultation with the supplier, select a combination of cleaners for trial in the plant. Option 2: Send one to three fouled elements to specialist for cleaning study, foulant analysis and review of plant performance history and pretreatment process. Document the findings along with pilot cleaning results using a recommended improved cleaning procedure. Simultaneously address recovery of the plant and avoidance of repeated fouling. Option 3: Send all fouled elements for off-site cleaning by specialist.

Strategy 3: All cleaning efforts by re-circulation of cleaning solutions have failed, consider non-routine methods like using proprietary membrane conditioning liquids or membrane reconstruction process by which membrane bundle is unrolled, cleaned leaf by leaf, then restored with a new hard-casing.

4.3. On-line Cleaning Procedure

Basically there are six (6) steps in the cleaning of membrane elements in place in RO systems:

1. Mix Cleaning Solution.

2. Low Flow Pumping. Pump preheated cleaning solution to the vessels at conditions of low flow rate (about half of that shown in Table 2) and low pressure to displace the process water. With the RO concentrate throttling valve completely open to minimize pressure during cleaning, use only enough pressure to compensate for the pressure drop from feed to concentrate. The pressure should be low enough that essentially no permeate is produced. A low pressure minimizes re-deposition of dirt on the membrane. Dump the concentrate, as necessary, to prevent the dilution of the cleaning solution.

3. Re-circulate. After the process water is displaced, cleaning solution will be present in the concentrate stream. Re-circulate the concentrate to the cleaning solution tank and allow the temperature to stabilize.

4. Soak. Turn the pump off and allow the elements to soak. Sometimes a soaking period of about 1 hour is sufficient. For difficult to clean foulants, an extended overnight soaking period of 10-15 hours is beneficial. To maintain a high temperature during an extended soaking period, use a slow re-circulation rate (about 10% of that shown in Table 2).

5. High Flow Pumping. Feed the cleaning solution at the rates shown in Table 2 for 30-60 minutes. The high cross-flow rate flushes out the foulants removed from the membrane surface by the cleaning, with minimal or no permeation through the membrane to avoid compacting the foulant. If the elements are heavily fouled (which should not be a normal

occurence), a flow rate which is 50% higher than shown in Table 1 may aid cleaning. At higher flow rates excessive pressure drop may be a problem. The maximum recommended pressure drop is 20 psi per element or 60 psi per multi-element vessel, whichever value is more limiting.

Note: In this cleaning mode, foulants are generally partially dissolved in the cleaner and partially dislodged physically from the membrane and flow channels without dissolving. An in-line filter removes the re-circulated particles, and should be monitored for cartridge replacement.

6. Flush Out. Pre-filtered raw water can be used for flushing out the cleaning solution, unless there will be corrosion problems such as with seawater corroding stainless steel piping. To prevent precipitation, the minimum flush temperature is 20 deg. C.

Additional Notes: The pH should be monitored during acid cleaning. The acid is consumed when it dissolves alkaline scales. If the pH increases more than 0.5 pH units, add more acid.

Feed Pressure*	**Element Diameter**	**Feed Flow Rate**
(psig)	**(inches)**	**Per Vessel (GPM)**
20 - 60	2.5	3 - 5
20 - 60	4	8 -10
20 - 60	6	16 -20
20 - 60	8	30 -40

* Dependent on the number of elements in the pressure vessel

Table 2. Recommended High Re-circulation Flow Rates During Cleaning.

Multi Stage Systems

For tapered multi-staged systems the flushing and soaking steps can be performed simultaneously in the entire array. The high flow-rate re-circulation step however should be carried out separately for each stage, so that the flow-rate is not too low in the first stage and too high in the last. This can be accomplished either by using one cleaning pump and operating one stage at a time, or using a separate cleaning pump for each stage.

4.4. Control and Improvement of Cleaning Process

To assure complete recovery of membrane performance by cleaning, the system performance should be adequately controlled by trending of normalized flux, differential pressure and salt rejection [20] to 1) trigger a cleaning when any monitored parameters change from normal baseline by 10-15%, 2) record the trended parameters before and after each cleaning, 3) initiate improvement actions for better cleaning if membrane performance does not fully recover. A change in responsiveness to previously effective cleaning process signals a change in fouling pattern that requires immediate attention. If partial cleanings are allowed to continue, the system performance will decline at increasing rate, and will become increasingly difficult to recover.

In-place cleaning processes are improved primarily by the choice of cleaning chemicals and the order of the application sequence. Depending on the composition of the complex foulants, when two or more cleaners are found necessary, often the order in which they are used is important. Also critical, but to lesser extents are the variables of time, temperature, and cross-flow rate.

Through thorough review of the water and pretreatment chemistry, analyses of the foulant composition and source, and customized selection of antiscalants, dispersants and high performance cleaners, both fouling avoidance and reliable plant performance can be attained. Practical experiences show that plant performances invariably change over years of service due to imposed changes in source water, equipment, regulatory and human factors. Attentive operation and maintenance assures early detection of developing problem, and timely adjustments.

5. Conclusion

High rejection of dissolved salts and suspended colloidal particles in RO feedwater cause scaling and colloidal fouling of membranes. Understanding the chemistry of membrane fouling and methods of control, coupled with keen monitoring during O&M, are necessary for the assurance of RO process stability. Ever increasing need for maximum water extraction, while reducing the volume of concentrate requiring disposal pose challenges to chemical understanding and control by O&M personnel. Information provided in this chapter provide key words and concepts for the readers to glean from the expansive literature.

For readers who assume responsibilities of existing RO plants, problems may have arisen due to inadequate pretreatment design, or due to changed sources of raw water. A companion chapter on pretreatment for reverse osmosis systems is available on line in an open-access book on Desalination [22].

Author details

Ph.D. Robert Y. Ning*

Address all correspondence to: Rning@kingleetech.com

King Lee Technologies, San Diego, California, United States of America

References

[1] Bergman, R. Ed (2007). Reverse Osmosis and Nanofiltration. , Manual M46, 2nd Edn., American Water Works Association, Denver, CO.

[2] Paul, D. H. (2003). The Four Most Common Problems in Membrane Water Treatment Today, Analyst, Winter, 1-5; and world-wide training seminars of D.H. Paul, Inc.

[3] Ning, R. Y. (2003). Discussion of Silica Speciation, Fouling, Control and Maximum Reduction. *Desalination*, 151, 67-73.

[4] Ning, R. Y., Troyer, T. L., & Tominello, R. S. (2009). Antiscalants for Near Complete Recovery of Water with Tandem RO Process. *Desalination and Water Treatment*, 9, 92-95.

[5] Ning, R. Y., & Netwig, J. P. (2002). Complete Elimination of Acid Injection in Reverse Osmosis Plants. *Desalination*, 143, 29-34.

[6] Ning, R. Y. (2001, September). Antiscalants That Permit Operation of RO Systems at High pH Levels. *Ultrapure Water*, 18-21.

[7] Ning, R. Y. (2000, September). Making an Ideal Process for Pure Water by Reverse Osmosis. *Water Conditioning & Purification*.

[8] Ning, R. Y. (1999, April). Operational Characteristics of Reverse Osmosis Process Chemistry. *Ultrapure Water*, 39-42.

[9] Ning, R. Y. (1999). Reverse Osmosis Process Chemistry Relevant to the Gulf. *Desalination*, 123, 157-164.

[10] Ning, R. Y., & Shen, P. T. L. (1998, April). Observations from Analysis of Reverse Osmosis Membrane Foulants. *Ultrapure Water*, 37-40.

[11] Ning, R. Y., & Stith, D. (1997, March). The Iron, Silica and Organic Polymer Triangle. *Ultrapure Water*, 30-32.

[12] Ning, R. Y. (2003). A Paradigm Shift from Prepurification of Membrane Feedstreams to Minimal Chemical Intervention. *Amer. Membr.Tech. Assoc. 2003 Annual Symposium, Westminster, CO*, August 3-5.

[13] Ning, R. Y. (2003). Sustaining the Production of Pure Water in Reverse Osmosis Plants for First Use or Reuse. *Annual Water Technologies Convention, Amer. Water Technologists, Phoenix, AZ*, September 17-20.

[14] Ning, R. Y., Troyer, T. L., & Tominello, R. S. (2003). Chemical Control of Colloidal Fouling of Reverse Osmosis Systems. World Congress on Desalination and Water Reuse, Intern. Desal. Assoc., Bahamas, Sept. 28-Oct.3; Desalination, 172, 1-6 (2005).

[15] Varnava, W., Silbernagel, M., Kuepper, T., & Miller, M. (1996, August). Reverse Osmosis Element Preservation Study. *Proceedings of Biennial Conf., Amer. Desalting Assoc.*, 308-327.

[16] Netwig, C. L., & Kronmiller, D. L. (1993). U.S. Patent No. 5,250,118.

[17] Ning, R. Y. (2010). Reactive Silica in Natural Waters- A Review. *Desalination and Water Treatment*, 21, 79-86.

[18] Ning, R. Y. (2009). Colloidal Iron and Manganese in Water Affecting RO Operation. *Desalination and Water Treatment*, 12, 162-168.

[19] Ning, R. Y., & Troyer, T. L. (2007). Colloidal Fouling of RO Membranes following MF/UF in the Reclamation of Municipal Wastewater. *Desalination*, 208, 232-237.

[20] Troyer, T. L., Tominello, R. S., & Ning, R. Y. (2006, September). A Method to Automate Normalization and Trending for RO Plant Operators. *Ultrapure Water*, 37-43.

[21] Ning, R. Y. (2003, September). Process Simplification Through the Use of Antiscalants and Antifoulants. *Ultrapure Water*, 17-20.

[22] Ning, R. Y. (2011). Pretreatment for Reverse Osmosis Systems, in Expanding Issues in Desalination, Ning, R.Y., Editor, InTech Open Access Book. Available: http://www.intechopen.com/articles/show/title/pretreatment-for-reverse-osmosis-systems.

Chapter 5

Standardized Data and Trending for RO Plant Operators

Thomas L. Troyer, Roger S. Tominello and Robert Y. Ning

Additional information is available at the end of the chapter

http://dx.doi.org/10.5772/49994

1. Introduction: Necessary Practical Knowledge to Manage Membrane Systems

This paper is written to strengthen the management of membrane facilities. Day-to-day operation of most large systems is done in a systematic and effective manner. However, management of membrane system resources over the middle to long term has not been so well addressed. For example, we find that the on-site operators are quite capable in maintaining high pressure pumps, detecting sudden rises in salt passage, or rapid feed pressure up ramps, but are not often able to determine if the membrane elements are experiencing a slow loss of productivity, or why a drop in second stage differential pressure needs to be addressed.

This problem in dealing with medium to long term issues is due not to individual operator failings, but to a systematic lack of training and support. Operators are not given sufficient information about membrane technology to know what the system is telling them. But they are often 'on the spot' to make crucial decisions.

Fortunately for operators, there are tools available which can assist them in operating and managing membrane systems. Using these, operators can educate themselves on the job.

1.1. Managing versus operating membrane systems

The job of an operator in a membrane system facility can be quite difficult. Typically they are responsible for maintaining not only the membrane system, but also feed water sources such as wells, water transmission, and disposal of concentrate waste streams, the normal building maintenance, regulatory compliance record keeping, and the list goes on. It is thus understandable the operator's focus is on the here and now.

Yet many important tasks necessary for the proper operation of the facility require the collection, and interpretation of system data over months and years. Some of these tasks include system optimization, cleanings, and membrane replacement. Each of these tasks can be carried out in a straight forward fashion by operators, but only if they are given the requisite tools and the time to learn to use them.

1.2. Managing membrane system resources

The following list of tasks are needed to operate a membrane system over the medium and long term. They require the proper use of membrane system resources such as labor, system data, chemicals, and membranes. All or part of organizing these resources typically falls to system operators to accomplish.

Each of the listed tasks below can be planned and timed using information derived from the analysis of system data.

1.2.1. Optimizing day to day membrane system performance

Optimization of system performance involves adjusting and tweaking operating settings of existing membrane equipment to best achieve the facility management goals. These goals may be producing product water at the lowest cost per gallon, using membrane technology to produce a potable product water from a feed water containing a contaminate such as nitrate, operating the system to minimize waste water disposal, etc.

To carry out such optimizations requires understanding of how the system is currently running, and how it runs after making changes. Looking at system data using the right tools is the only way operators can carry out this task with a high degree of confidence.

1.2.2. Optimizing membrane cleanings

In the course of operating membrane systems some foulant will accumulate on the membrane surface reducing permeate flow. At some point the operators have to decide when to take the system off line and clean to restore productivity. If they clean too soon, the membranes are overexposed to cleaning chemicals. This can lead to shorter membrane lifetime. If they wait too long, the foulant may be very difficult to remove. Using the right tools allows operators to forecast from system data when to clean. In this fashion membrane cleanings are optimized such that the time between cleanings is maximized and the actual time to clean the system is minimized.

http://dx.doi.org/10.5772/49994

1.2.3. Optimizing membrane change out

Membranes have a life time after which they have to be changed out and new membranes installed. They may fail due to degraded salt rejection, low productivity, or high differential pressure. They may no longer reject a particular contaminate and thus produce permeate water not meeting potable water standards.

As replacing a load of membranes is expensive and time consuming, it is almost mandatory that the event be forecast as far in the future as possible as can be done. Data analysis is required to make such forecasts.

1.2.4. Pilot Tests

Many facilities have small membrane pilot test system or frequently have such equipment bought in by vendors. The most common use of these pilot systems is to qualify new membranes made by different membrane manufacturers for use in the system. Other tests would include testing elements taken from the current load of elements to better qualify performance, or testing different operating conditions considered too risky to trial on the main system.

Proper analysis of the data from the pilot test as it is being run, and after the test is completed, is necessary for the test to have value.

1.2.5. Pathologies/Troubleshooting

As with all processes, there are occasional upsets, mistakes, or the unforeseen. The effects are usually obvious such as low productivity or low salt rejection. But the cause of the trouble cannot always be determined directly from reading the gauges or looking at the SCADA screen. Nor is it always obvious where to dig into the equipment to look for answers.

As we will discuss below, the answers for membrane system problems lie in analyzing how the system has operated over time.

1.2.6. System equipment modifications/upgrades

For some unlucky operators, the membrane system itself is poorly designed, or has been retasked. Or the system has to be upgraded to meet new governmental or market demands. While in such circumstances facility managers call in engineering resources, the operators are also involved, especially if they have been collecting and analyzing the system data. In many cases it is the operators who may have initially identified the system deficiencies.

If operators have the proper tools and experience they can make significant contributions to retooling their systems.

2. Membrane System Variables and Interrelationships

To carry out the tasks listed above requires the operators to collect and analyze the operating data available on most membrane systems. Medium and larger membrane systems are organized into multiple stages, usually two stages. See Figure 1. Each stage has three process streams; feed, permeate and concentrate. Each stream is characterized by pressure, flow, temperature, and conductivity. So there are many variables available to record and analyze.

To further complicate data collection and analysis, the value of each of the variables is strongly dependent on the values of the other variables. For example as feed pressure increases, permeate and concentrate flows can increase and permeate conductivity decreases. As temperature increases, flow increases, permeate conductivity increases, feed pressure can decrease [1].

Without assistance, a system operator would be hard pressed to accomplish any task that required them to utilize the system data they have collected. Fortunately there are many easy to use computer tools available so that the operator need only collect the indicated data, enter it, and click a button on screen to get the required analysis.

All of these computer tools are based on a written procedure, #4516, distributed by the ASTM organization [2]. The ASTM procedure consists of a set of equations into which system data can be inserted and calculations performed. To avoid having to do the calculations by hand the equations have been put into Excel spreadsheets [3-5]. The added benefit of using the spreadsheet is the ease of creating graphs from the calculated values. These graphs are far easier to interpret than a table of numbers.

For this paper, we are going to use an automated spreadsheet program called System Wizard running under Microsoft Excel 2000 [6]. See references. Using System Wizard we will show how to carry out the tasks mentioned above.

3. System Wizard, an Overview

As discussed above the performance of membrane systems cannot be directly measured by following a single variable. Performance can only be seen after the system data has been analyzed. To carry out the analysis you need to be able to directly compare one day's data to another. But membrane system data can vary from day to day. Pressure can go up, and simultaneously temperatures go down. These changes affect the product flow and salt reject. The changes have to be sorted out. Thus it is necessary to create calculated values that allow for these interactions so that each days data can be standardized for direct comparison. Once we do that, we can graph the calculated values, look for system performance trends, and make well founded system performance evaluations.

http://dx.doi.org/10.5772/49994

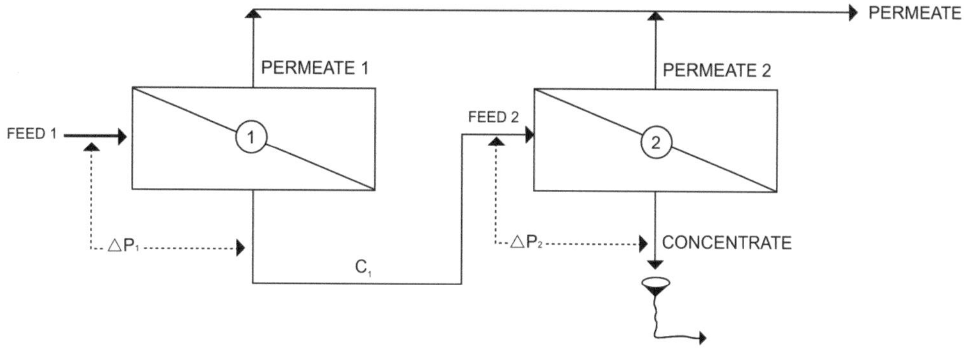

Figure 1. Medium and larger membrane systems are organized into multiple stages, usually two stages.

To use an analogy, if we wish to see how well our automobile is performing we may calculate the vehicle miles per gallon, mpg, after each fill up. To determine performance over the lifetime of the car, we can graph the mpg versus time to see how performance varies over time, season.

We can then use our mpg information to make decisions. For example if currently mpg is 27.3 and the new car mpg was 33.2 we can decide to get a tune up. After the tune up, we can see from the 'tuned up' mpg whether the work on the car was effective. We may look over previous mpg versus tune ups and see if it is time to start looking for a new car.

To use membrane system data to determine system performance we need to calculate values that cancel out variations in pressure, temperature, flow and conductivities. While there a great many ways of going about this [7], the ASTM organization has reviewed the membrane industry practice and published procedure #4516 that is now widely used.

The ASTM procedure calculates 'standardized' permeate flow and salt rejection values from daily system data. These calculated values are derived from equations that compare each of the system variables to a standard set of variables.

For example, suppose we use for standard values the values measured on the first day of operation of a new RO membrane system. Then we operate the system for a month. How as the system performance changed? Looking at Table 1, Day 30, Case 1 we see that the all the system values remained the same save permeate flow which has decreased from 1000 to 500 gpm.

How would we rate the system performance at 30 days? On the basis of permeate productivity we can easily see that the system is producing 50% less product even though the pressure, temperature, etc., are unchanged. So we see that the system performance has decreased by 50%.

In Case 1, it is easy to rate the productivity. Only the permeate flow has changed. However, if we look at Day 30, Case 2 we see that while the permeate flow and other values have not changed, the Feed and Concentrate Pressure has doubled. Has the permeate productivity performance changed?

System Variables	Day 1 Standard	Day 30 Case 1	Day 30 Case 2	Day 30 Case 3
Feed Pressure, psig	100	100	**200**	156
Permeate Pressure	0	0	0	2
Concentrate Pressure	90	90	**180**	146
Temperature, ºC	25	25	25	33
Permeate Flow, gpm	1000	**500**	1000	2000
Concentrate Flow	333	333	333	666
Feed Conductivity ,µS	500	500	500	689
Permeate Conductivity	5	5	5	5
Standardized Flow	**1000**	**492**	**479**	**998**
Flow Performance, %	**100**	**49.2**	**47.9**	**99.8**
Standardized Salt Rejection	**99.0**	**99.4**	**99.0**	**98.9**
Rejection Performance, %	**100**	**100.4**	**100**	**99.9**

Table 1. System Data and Standardized Data.

Yes, since it now takes twice the pressure to make the same amount of permeate. We see that again the system performance has decreased by 50%.

In Case 2, as in Case 1, the difference between the data of Day 1 and Day 30 was rather easy to see. The Feed and Concentrate pressure was doubled, but the permeate flow did not increase. In Day 30, Case 3, all the system values are different from Day 1. It is not possible to 'eyeball' this data set and say how the permeate productivity performance has changed. The pressures have gone up, the temperature has gone up, the flows have gone up, and the feed conductivity has gone up. This is where it is necessary to use the ASTM procedure to determine system performance.

Using System Wizard (that is based on the ASTM procedure) we can calculate the standardized values for each case. We enter the values of the variables listed in Table 1 and then click the on-screen button to calculate the standardized values. The results are shown in the last rows of Table 1. Here we have used the Day 1 data as the standard to compare against the three cases. As shown in Table 1 'Flow Performance%', we see that Case 1 and Case 2 permeate productivity is approximately 50% of the Day 1, while Case 3 productivity is virtually the same. We also calculated the standardized salt rejection for each case.

From these examples it is clear that standardized values are required if we are to be able to compare one days system performance to another days.

http://dx.doi.org/10.5772/49994

4. Monitoring membrane system performance using System Wizard

Just as we may monitor our automobile and make decisions using mpg, we can use Standardized Flow and Salt Rejection to oversee membrane systems. Using System Wizard or similar programs we can evaluation system performance and best deploy facility resources including labor, chemicals and membranes.

4.1. Optimizing day-to-day membrane system performance

We have discussed calculating standardized values from system data. See Table 1. However, in practice we do not attempt to monitor system performance using a table of values. Instead we graph the standardized values versus time. The graphic format allows us to 'eyeball' a large number of data points so we can find trends in the data. Figure 2 shows a graph of the standardized values for a RO system. From the level trend of the data we can easily determine that this system's permeate productivity is quite constant. The salt rejection is also seen to be quite constant when we learn that the RO feed water alternates between two wells explaining the up and down grouping of the data points.

We can again see the stability of this RO system when we look at Figure 3. Here we graph the differential pressure of the system versus time. Again we see a level trend.

The system shown in Figure 2 and 3 operates on very good quality feed water. The trends are what all membrane systems would have ideally. However, it is more common that the feed water has significant fouling material. Where that is the case the productivity trends downward as shown in Figure 4. In this case, the effect of the foulant layer is to reduce the productivity by approximately 10% over the course of 17 months (from 6/21/11 to shortly before 01/07/12). For a large membrane system this decline is quite acceptable.

We see that trending standardized system data can tell us how well our membrane system is operating. We can take this procedure to a new level by making operation changes and following the effects.

4.1.1. Membrane Flux optimization

Membrane flux is the expressed as gallons of permeate per square foot of membrane per day. For the case where the feed flow is increased and the recovery is unchanged, the higher the flux is raised, the higher the total productivity of the system. However, as the flux goes up, the more likely that foulants will be drawn to the surface of the membrane which would reduce productivity.

By raising the flux in a step-wise fashion over a reasonable period of time and trending the productivity we can determine the point at which the productivity is maximized. What we would see when the data is graphed is the productivity trend would be relatively flat as seen in Figure 4 as the flux was increased. But then the productivity starts to slope down in a noticeable fashion as we pass the optimum flux and a faster rate of fouling is established. From these results we can determine the optiminum flux.

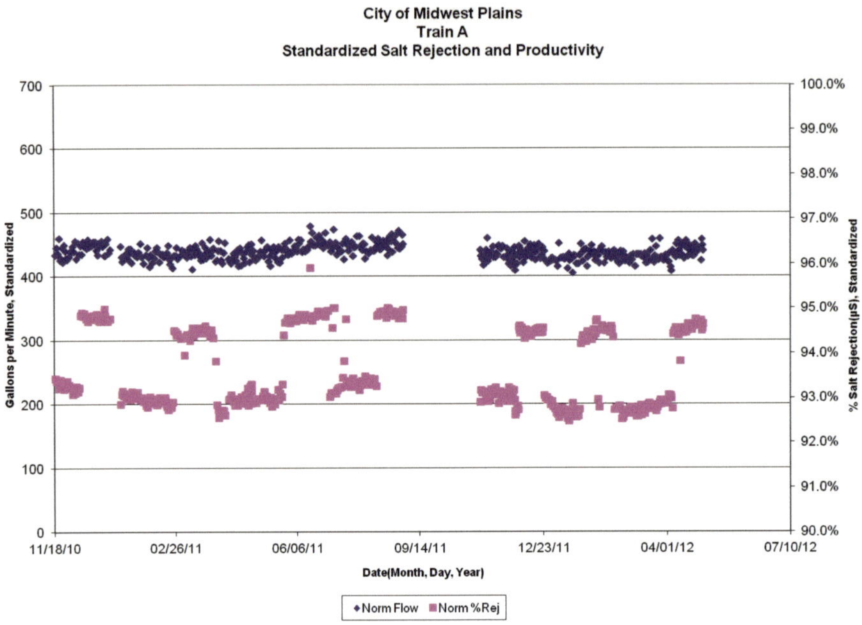

Figure 2. Standardized System Data: Permeate Productivity and Salt Rejection.

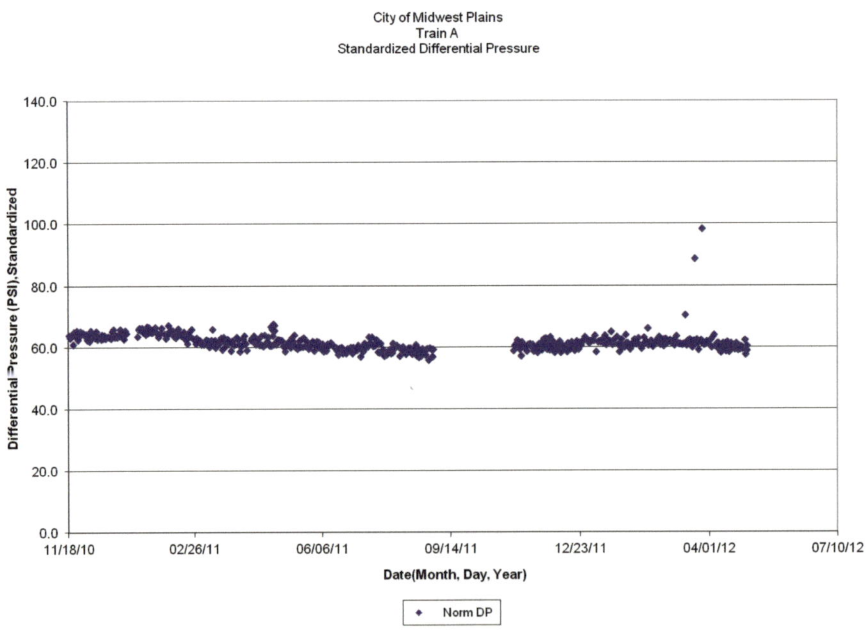

Figure 3. Standardized System Data; Differential Pressure.

http://dx.doi.org/10.5772/49994

4.1.2. Recovery optimization

Recovery is the percentage of permeate water produced from feed water. Recovery can range from 30% to 90% depending on quality and cost of the feed water and cost of concentrate disposal. For the case where the feed flow is kept constant but the permeate flow increases and the concentrate flow deceases, the recovery of the system increases resulting in higher productivity. Under these conditions the bulk flow (average flow of water through the system from feed inlet to concentrate outlet) decreases since more water is flowing through the membrane. The slower the bulk flow does not sweep the surface of the membrane as well as the original higher bulk flow and can result in a higher fouling rate.

Additionally, the flux also has also increased which can lead to faster fouling. When the standardized productivity is graphed versus time we would see initially a relatively flat trend. As we step up the recovery over a few weeks or months there would come a point where the recovery is high enough that the fouling rate affects the productivity and it starts to trend downward a rapid rate. From these results we can determine the optimum recovery.

4.1.3. Other Optimizations

As discussed above, the effect of any membrane system change can be evaluated by examining the resulting productivity, salt rejection and differential pressure trends. For example, besides changes in membrane flux, and recovery, operators can determine the effect applying permeate back pressure to balance flows from individual stages of multi-stage systems.Another use of this procedure is the very common situation in which the feed water is affected by introducing a new source such a newly dug well. By trending the standardize system values it is possible to see the effects the changed feed water and see if the current blend of source waters needs to be adjusted. It is also possible to determine if additional pretreatment is needed.

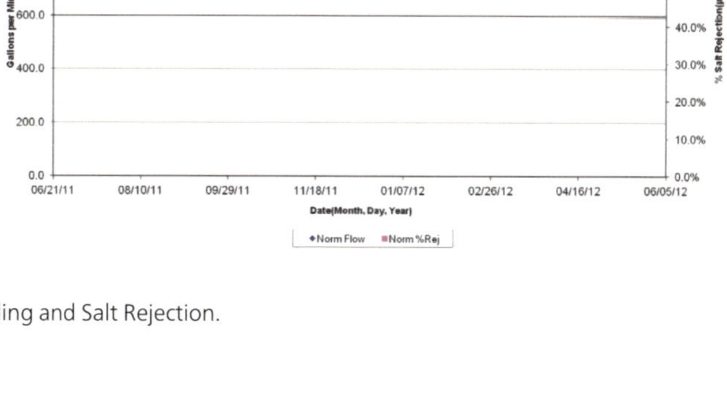

Figure 4. Normal Fouling and Salt Rejection.

4.2. Membrane Cleanings

All membrane systems eventually need to be cleaned. Knowing when to clean, and deciding if the cleaning was successful, are key decisions operators must make. Waiting too long to clean can result in having so much foulant on the membrane that it is very difficult to remove. Cleaning too soon exposes the membrane to harsh chemicals too frequently which can lead to chemical aging of the membrane. A good rule of thumb is to clean membrane systems when the productivity has decreased by 15% since the last successful cleaning. This rule can be modified based on the experience of the operators.

An instance of membrane cleaning is shown in Figure 5. The system was put on line May 29th, 2011. The initial standardized productivity on that date was 1206 gpm. On October 21st, after 292 days in service, the operators took the system off line to be cleaned. On that date the standardized productivity was 1073 gpm, a decline of 11% as compared to May 29th. After cleaning the system was put back on line November 1st. The standardized productivity for that date was 1197 gpm. Since this productivity is 99.3% of that of May 29^{th,} we conclude that the cleaning was successful as it restored the productivity to the same level seen on May 29th.

Taking our procedure to the next level we can use the time between cleanings as another measure of system performance. Since we are determining when to clean based on the data we can assume that 'cleaning period', the time between cleanings, should be constant. If we change the operation of the system in some fashion we can evaluate the change on how the cleaning period is affected.

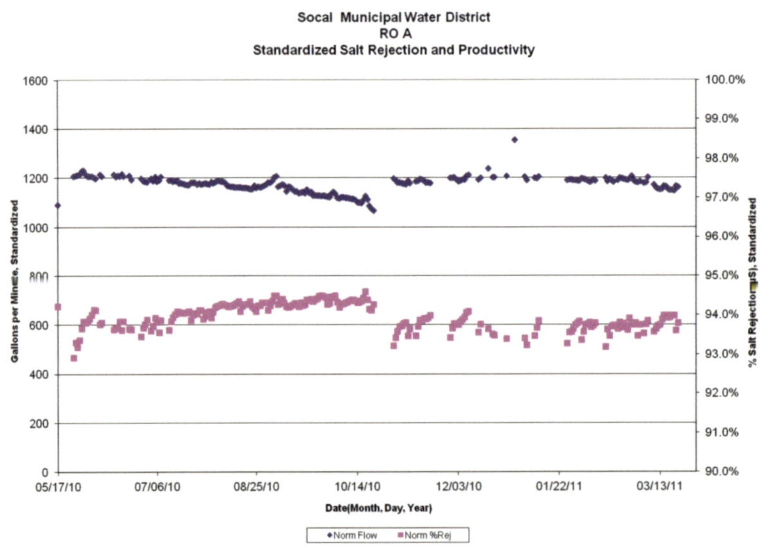

Figure 5. Trending Membrane Cleaning.

http://dx.doi.org/10.5772/49994

4.3. Membrane change out

Membranes are changed out due to poor rejection, poor productivity or high differential pressure. In some cases, the change out is due to a sudden event such as an accidental exposure of the membrane to an oxidant. But in the remaining instances a membrane change out is due to normal wear and tear over time. In these cases it is possible to forecast when a change out will occur using trends from standardized system data.

An example of such a forecast would be trending declining salt rejection. After a set of elements has been cleaned a number of times over the years, the salt rejection can start to deteriorate due to chemical aging and irreversible fouling. By comparing the salt rejection after the last cleaning to the previous cleanings, or graphing such values over time, salt rejection decline can be trended. Operators can predict when the membranes will need to be changed out by extrapolating the trend into the future to the point at which the salt rejection falls below the systems product water specification.

4.4. Pilot tests

Some membrane facilities have pilot test equipment in which individual membrane elements can be tested. Pilot test equipment can be used to answer simple questions such as does a particular element have low productivity. However, when the questions are more complex, the data generated from the test must be properly analyzed.

Operators can use the standardized data approach in pilot tests by first running the test element under a set of baseline conditions for reasonable period of time to establish a trend. Then the element can be run under the test conditions and that data trended. The baseline and test trends can then be compared and evaluated.

Another use of pilot test equipment is to qualify new types of membranes. Using standardized data allows direct and unambiguous comparison.

4.5. Pathologies

Membrane systems without problems have straight level trend lines similar to Figure 2 and 3. Any deviations are due to the usual operating problems such as scaling. For example we can see in Figure 5 the effect of fouling on the productivity trend line. However, when unusual situations arise, they can also have a particular effect on trend lines.

4.5.1. Low Bulk Flow

Bulk flow is the average flow through a stage of a membrane system. For example, a second stage containing 10 pressure vessels may receive 400 gpm and recover 200 gallons as permeate, thus 200 gpm of concentrate exits the stage. The bulk flow is the average of 400 gallons in and 200 gallons out, and would be equal to 300 gpm.

Maintaining sufficient bulk flow is important. Bulk flow sweeps the surface of the membrane keeping foulants in suspension and pushing them out of the stage. If the bulk flow drops too low, foulants can accumulate on the membrane surface and rapidly reduce permeate flow.

Bulk flow can be reduced for example if the stage recovery is too high, or the stage receives too little feed water from the upstream membrane stages. However, knowing when bulk flow is problematic can be difficult to determine before the system has actually started. Site specific factors may lead to operating conditions the system designers did not anticipate.

Trending the differential pressure (dp) of each of the systems stages can be used to diagnose bulk flow problems. Figure 6 is a graph of the first and second stage differential pressures versus time. Approximately July 2006 the system was cleaned. Both the stage dps immediately start trending down. That indicates that the bulk flow in each stage is decreasing since differential pressure, inlet pressure minus outlet pressure, is directly proportional to bulk flow. As bulk flow goes down, differential pressure goes down.

The reason that the first stage dp is going down is that the second stage is being rapidly fouled and losing productivity. To make up this loss in productivity the first stage must make more permeate which lowers the amount of water exiting the first stage and going to the second stage. Remember that bulk flow is the average of the inlet and outlet flow, thus both stage dps must go down.

An example is shown in Table 2 below. You can see that the inlet to the first stage and outlet from the second stage are constant, meaning the recovery is constant. Initially, the first stage makes 200 units of permeate and the second stage makes 100 units.

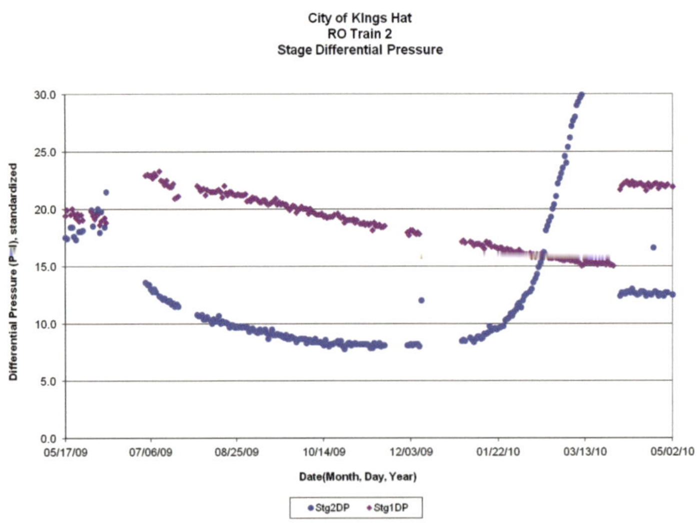

Figure 6. Low Bulk Flow.

Later, the second stage is only making 50 units. This forces the first stage to increase permeate flow to 250 units. Consequently, the bulk flows for both stages goes down and so do the respective stage dps.

Stages			**First Stage**				**Second Stage**	
Flows	Inlet	Outlet	**Permeate**	**Bulk Flow**	Inlet	Outlet	**Permeate**	**Bulk Flow**
Initial	400	200	**200**	**300**	200	100	**100**	**150**
Later	400	150	**250**	**175**	150	100	**50**	**125**

Table 2. Example of Data for Initial and Later Bulk Flows.

Going back to Figure 6, we can see after December 2009, the second stage dp starts to rise and continue to rise dramatically going from 10 to 60 psi. This occurs as the second stage fouling starts to fill in the feed spacer and restricts flow. The effect is the same as putting a thumb over the end of a garden hose. This is a very serious condition and requires immediate attention.

4.5.2. Particle Fouling

Particle fouling occurs when solid debris passes through the systems macroscopic pretreatment filtration, enters the membrane system and starts to fill the membrane element feed spacer. The first stage of the system of the system is affected and the first stage differential pressure begins to rise. Graphically the first stage dp is trending up while the second stage dp remains unchanged.

As an example of a sudden increase in first stage dp there might be a well casing failure. This eventual will be spotted from normal gauge readings. However, if the stage dp data are collected and trended it is often possible see the increase very early on and potentially to step in and correct whatever malfunction is the cause.

4.6. System Equipment Modification

For some unlucky operators, the membrane system itself is poorly designed, or has been retasked. Poor or inadequate design is more common for smaller membrane systems but larger systems also can have issues. Using standardized data to trend system performance allows operators to see that the problems they have to deal with are not some new factor that developed over time, but began at systems initial startup. For example, if the system was designed for an optimistically high permeate flux, the fouling rate of the system may be too high. In this case the operators may be faced with very frequent membrane cleanings. Another example would be poorer than expected feed water quality. This again may lead to very frequent cleanings. Yet another case would be an imbalance between the system high pressure pumps and the membranes loaded in the system. This has been seen where the pumps cannot supply sufficient flow to the second or third stages of a membrane train. There are many ways that a new system or a newly re-tasked system can run into difficulties. Trending standardized data from initial start up not only identifies but also documents design problems which can be important when dealing with contractual performance guaranties and warranties.

5. Summary and Conclusion

Standardizing and trending data using available computer tools enables operators to carry out tasks such as system optimization, membrane cleaning, membrane replacement, etc., as part of their daily routine in an effective and efficient manner.

Currently, most operators are not using these tools to assist them to manage their membrane systems. Consequently membrane systems are not managed as efficiently as they could be.

Using available tools to standardize and trend system data would educate operators on the job on how membrane systems actual work, what to look for, and how to make good decisions.

Supervisors and managers of membrane systems who want to optimize overall operation of their facilities can move forward by encouraging the adoption of standardization and trending by operators.

Author details

Thomas L. Troyer*, Roger S. Tominello and Robert Y. Ning

*Address all correspondence to: tomtroyer@kingleetech.com

Technical Service Group, King Lee Technologies, USA

References

[1] Byrne, W. (1995). Reverse Osmosis A Practical Guide For Industrial Users. Littleton: Tall Oak Publishing.

[2] ASTM (2010). Volume 11.02 Water (II) #4516-00 Standard Practice for Standardizing Reverse Osmosis Performance Data. Conshohocken: ASTM, 2010.

[3] Hydranautics ROdata download. http://membranes.com/index.php?pagename=rodata (accessed May 17, 2012).

[4] FilmTec FTNorm download. http://www.dowwaterandprocess.com/support_training/design_tools/ftnorm.htm (accessed May 17, 2012).

[5] TorayUSA TorayTrak download. https://ap8.toray.co.jp/toraywater/webapp/client/softwareSelect.do (accessed May 17, 2012).

[6] King Lee Technologies System Wizard download. www.kingleetech.com.

[7] Zhao, Y., & Taylor, J. (2005). Assessment of ASTM D 4516 for evaluation of reverse osmosis membrane performance. . Desalination 2005; 180 , 231-244.

Section 3

Selective Waste Removal

Chapter 6

Processing of Desalination Reject Brine for Optimization of Process Efficiency, Cost Effectiveness and Environmental Safety

M. Gamal Khedr

Additional information is available at the end of the chapter

http://dx.doi.org/10.5772/50234

1. Introduction

Reverse Osmosis (RO) is currently confirmed and generally approved as the most feasible technology for desalination of brackish groundwater being the most economic for its range of salinity over a wide range of production capacities, and in view of its lowest requirements of energy, and its application ease.

The currently acceptable norm of recovery of desalted water in projects of brackish water reverse osmosis (BWRO) ranges usually between 65 to 85 % according to raw water quality, level of chemical pretreatment and concept of plant design/operation, would it be intended to be a sophisticated facility of low operation cost or vice versa. The balance of 15 %, or above, the desalination reject stream in which the RO rejected components are concentrated, is disposed as a wastewater (WW). Among the disposal options selected to get rid of the desalination reject stream are: 1) Sewer stream, 2) Land application including percolation, 3) Deep well injection and, 4) Evaporation ponds. The last option is the most common in the Middle East in view of:

- The rather common high temperature
- The low ambient humidity
- The relatively low cost of land in desert areas

Disposal of RO reject water aims, in most of the alternatives, to just get rid of that stream without further water recovery which wastes the cost of initial pumping and chemical treatment. It is, therefore, evident that the increase of desalted water recovery is a main factor in determining the process cost effectiveness. On the other hand, a too high recovery would lead to most, if not all, the membrane fouling problems and the subsequent decline of performance and eventually membrane damage [1]. The present work investigates the promotion of the RO desalination efficiency and cost effectiveness.

Desalination reject stream (DRS) represents, in fact, a WW disposal problem. It includes, in addition to increased salinity, higher concentrations of polyvalent ionic species [2] due to the preferential high rejection of e.g. hardness components, heavy metal cations (HMCs) [3] or radioactive isotopes [4], organics [5]. DRS includes also the residual pretreatment chemicals of the primary desalination step, i.e., coagulants as iron or aluminum salts or polyelectrolytes, disinfection by products, antiscalants [6].

In big RO desalination facilities, however, the surface area of evaporation ponds may attain several millions of square meters and represents, therefore, one of the main cost factors of the desalination projects [7] due to the cost of land and of installation of ponds, digging, lining, construction of dykes [8].

Besides the considerable cost of installation of evaporation ponds and their annual maintenance, they may cause considerable environmental threat through:

1. Possible leak of concentrated brine and possibly contaminated water to pollute the groundwater reserves.

2. Flooding of ponds which was reported for many existing desalination plants in view of inadequate initial design or operation problems. Flooding of contaminated reject would contaminate the neighboring habitat.

In view of the increasingly stringent environmental regulations related to disposal of WWs and the high cost of evaporation ponds the present laboratory and pilot investigation work aims to promote RO desalted water recovery and reduction of the disposed brine stream to a minimum value so as to realize:

1. Promotion of the desalination process efficiency and saving of groundwater reserves.

2. Saving of, or lowering the cost of installation and maintenance of evaporation ponds.

3. Conformity with environmental regulations of WW disposal.

4. Reducing environmental risks of pollution of groundwaters.

Processing of DRS is supported by:

1. The progressive development of water treatment chemicals as the introduction of antiscalant of specific action as e.g. SiO_2or SO_4^{2-} specific antiscalants which enable safe operation of RO at higher recoveries despite the presence of higher concentrations of the scale forming components.

2. The creation of new generations of RO and NF membranes according to the trends of:

   a. Higher salt rejection

   b. Lower energy consumption

c. Optimized hydrophilic/hydrophobic characters and fouling resistance

3. The recent introduction of sensitive energy recovery systems capable of recovery of the residual pressure from the BWRO reject stream.

Our previous results of desalination reject processing through laboratory and pilot investigation [7] showed remarkable optimization of recovery of desalted water which increased the total RO process recovery up to 95 %.

Comparative evaluation of performance and cost of several alternatives of brine processing was conducted as e.g. application of high rejection, low energy, secondary RO together with use of specific antiscalants or partial softening NF of reject stream prior to secondary RO [7]. A primary cost analysis showed that the studied reject processing is quite cost effective even without consideration of the reduced surface area of evaporation ponds and consequently their cost.

Superior rejection of polyvalent cations from the reject stream was observed by NF as compared to hot lime softening (HLS) together with absence of chemical dosing stoichiometric to deposition of hardness components and, consequently, absence of sludge formation which represents itself a daily disposal problem. NF, on the contrary of HLS, does not require subsequent filtration. NF also leads to partial desalination of the brine stream while HLS results in increase of the concentration of some components like sodium and carbonate ions, and does not modify other components not included in the softening reactions as SO_4^{2-} and Cl^- ions and, therefore, results in increase of total dissolved solids (TDS).

As for the reject streams,where radioactive isotopes and/or HMCs were concentrated upon primary BWRO, treatment by NF and low energy RO revealed, under adequate application conditions, more efficient than conventional methods of WWs treatment [4]. In fact, several technical challenges remain with regards to the efficiency and cost of conventional methods for rejection of these contaminants. NF and Low energy reverse osmosis (LERO) were evaluated in this respect in comparison with methods of chemical precipitation, chelating ion exchange resins (IER's), hot lime softening and coagulation/settling/precipitation. Membrane methods gave higher rejection of radionuclides and HMC's ranging from zero to 20 pCi/L could match the maximum contaminant level (MCL) of the US Environmental Protection Agency (US-EPA) for drinking water. NF and LERO, on the contrary of the other methods, are continuous processes which are not shutdown for regeneration, do not suffer from interference of similar valence ions with contaminant separation and are not limited by high pH dependence.

Investigation of desalination reject processing (DRP) is of economic and strategic interests in view of the huge daily production rate of such stream. In Riyadh region alone, according to data from the National Water Company [9], if the main desalination facilities, Wasiea, Buwaib, Salboukh, Manfouha, and Malaz are operated at original design rate, the yearly rate of reject stream amounts to 30 million m^3/year which is expected to increase to more than 45 million m^3/year upon installation of the new Wasia project. It is, therefore, expected that the total BWRO reject in KSA, in view of the planned giant projects in Ha'il, Tabuk, etc..., would amount to > hundred million m^3/year.

2. Literature survey

Processing of brine concentrate of water desalination has been conducted for various purposes. For salt extraction, Sommariva et al [10] Smith and Humphreys [11] considered the processing of the desalination reject up to zero discharge using concentrate disposal processes among which solar/evaporation ponds until crystallisation. They stated that evaporation ponds are preferred in presence of strong solar radiation, low precipitation, and low cost desert land. Produced salts were proposed for use in agriculture, forestry, fauna, and algae production, and energy production. Ahmed et al [12] investigated salt production from reject brine by SAL-PROC technology which consists in multiple evaporation and/or cooling steps.

For the purpose of environmental protection, Shahalam [13] evaluated the removal of nitrogen and phosphorus from RO reject of refining effluent of biological processes treating municipal WW. While RO is proven to be effective in producing high quality effluent water for non potable uses e.g. for irrigation purposes, its reject contains too high amounts of P and N compounds harmful for the environment if the feed to RO units is effluent stream from municipal and industrial WW treatment plants. Brine treatment included activated sludge treatment and then granular medium filtration. Heijman et al [14] considered the pretreatment of RO and NF reject so as to attain recovery as high as 99% aiming to overcome the problem of reject disposal. A complicated and expensive sequence of steps is proposed and pilot tested that consisted of precipitation of hardness components at high pH, sedimentation, cation exchange resin, and then NF. As for desalination by NF or RO of surface water a more complex processing was tested, i.e. cation exchange resin, then Ultrafiltration (UF), NF followed by treatment by granular activated carbon (GAC). A recovery of 97% was achieved. For a still high recovery up to 99% SiO_2 removal was conducted by co-precipitation with Mg hydroxide at high pH. The total treatment scheme included double barrier against pathogens (UF and NF) and against micropolutants (NF and GAC). Furthermore, the resulting suspended particle concentration is low and the biological stability is expected to be excellent.

According to Jeppesen et al [15] disposal of highly concentrated brines poses significant environment risks. Extraction of some metals from this stream can multiple environmental and economic benefits. Removal of P has little economic benefit but may become interesting in view of environmental restrictions. This study showed that recovery of NaCl from brine can significantly lower the cost of potable water production if employed in conjunction with thermal processing systems. The high ammonia, sulphate, TDS and HMC's render the RO brine hazardous if dumped untreated [16]. Denitrification of RO brine concentrate was conducted by Anaerobic Fluidized Bed Biofilm Reactors with GAC media.

The main purpose for most of the research work related to reject processing was to promote the desalted water recovery by various techniques. Queen et al [17] treated the RO brine by NF for removal of polyvalent cations then it goes to the concentrate compartment of an Electrodeionization unit (EDI) while the initial RO permeate goes to the diluate compartment. Overall consumption of feed water was, therefore, reduced. However, on site reject treatment by EDI was reported to be effective only for small RO treatment units [18], while for large reject stream rates the cost can be very high. A modified evaporation system that consists of forced air thermal evaporation using turbine technology so as to create a high wind speed and generate a very high air temperature was used. This system is approved by US-EPA. It can operate in high humidity, low temperature conditions, and can evaporate up to 126 GPM at the cost of just discharge to sanitary sewer. Evaporation of RO reject was also investigated in underground rock salt mining operation [19].

In case of inland communities which have no ready sink for RO brine the disposal cost will increase significantly the cost of RO treatment, specially with the limited recovery to avoid scale deposition. Coral et al [20] studied the minimization of RO reject through vibratory shear enhanced process (VSEP) without softening. They stated that strategies to minimize brine volume include 1) pre RO softening to remove hardening components and achieve higher recovery, 2) two stage RO interrupted by brine softening, 3) innovative technologies for extraction of water from RO brine without softening e.g. VSEP. The same technology was used by Arnold [21] for optimization of water recovery from RO brine issued by Central Arizona Project and by Cates et al [22], in both cases, however, no cost analysis was conducted and no justification was given for selection of such expensive technique.

Electrodialysis (ED) [23] was also applied for treatment of brine resulting from RO treatment of textile effluent for the purpose of reduction of TDS with the recovery of acids and bases. The WW of textile dyeing was first treated by coagulation/precipitation for color removal followed by RO. RO reject was then treated by ED. This treatment was qualitatively reported to enable the protection of environment from contamination by dyes and the related additives and to promote the reject water recovery.

Capacitive deionization (CDI) was used for SWRO reject treatment [24] instead of blending the brine with secondary effluent and discharge to the sea. The objective of the project was to increase water recovery to more than 95% at the required water quality. Correspondingly the volume of the brine will be reduced to less than 5%. For inland RO facilities where disposal of untreated RO brine has adverse environmental impacts, this approach would represent a cost effective alternative for the management of the brine stream. Results of pilot testing have met expectation. Lee et al investigated the treatment of RO brine towards sustainable water reclamation practice [25]. RO brine generated from water reclamation contains high concentrations of organic and inorganic compounds. These authors concluded that *cost effective technologies for treatment of RO brine are still relatively unexplored.* The proposed treatment consists of biological activated carbon (BAC) column followed by CDI for organic and inorganic removal. 20% TOC was removed by BAC while 90% conductivity reduction was realized by CDI. Ozonation was used to improve the biodegradability of RO brine. The laboratory scale O_3 + BAC was able to achieve three times higher TOC removal compared to using BAC alone. Further processing with CDI was able to generate product water with better water quality than the RO feed water. The O_3 + BAC reduced better the fouling in the successive CDI step [26].

Duraflow Company [27] employed a three step approach to define a pretreatment process compatible with the recovery of RO brine using a secondary RO. The objective was to remove all components detrimental to secondary RO and realize suitable values of Silt Density Index (SDI).

The three step approach includes:

I. RO brine analysis to determine the components

II. Chemistry Development which is based on type & concentration of fouling substance identified in the RO brine, a chemical treatment process is developed to counteract each of the fouling factors:

1. Cold lime Softening
2. Colloidal silica removal by adsorption on $Mg(OH)_2$
3. Activated Carbon for organic reduction & oxidant destruction
4. pH optimization for the selected treatment & the secondary RO

III. Microfiltration to the adequate SDI then secondary RO

Kepke et al [28] considered the options of RO brine concentrate treatment:

1. Deep well injection
2. Natural treatment systems (Wetlands)
3. Electrodialysis/Electrodialysis reversal
4. VSEP membrane System
5. Precipitative softening/RO
6. High efficiency RO (pretreatment step [may be several] +secondary RO)
7. Mechanical evaporation
8. Evaporation ponds
9. Landfill

They defined high efficiency RO as a combination of the hardness removal pretreatment which include Lime soda softening followed by filtration and weak cation exchange resin.

This type of RO treatment is relatively new. It has not been used for water reuse applications but has been applied in the power stations and mining industries. The advantages of this process over Conventional RO include reduction in scaling, elimination/reduction of biological and organic fouling due to high pH where SiO_2 solubility is high. The expected recovery would attain 95%.

IER's were also applied in desalination brine reclaim. This did not only optimize system efficiency through additional permeate recovery but also reduced the amount of water and salt required for softener resin regeneration. Some of the salt in the last part of the brine cycle is used for the next regeneration of the exhausted resin.

http://dx.doi.org/10.5772/50234

According to the survey report "Managing Water In The West" [29] by the Southern California Regional Brine-Concentrate Management, the concentrate disposal technologies include 1- the volume reducing and 2- the zero liquid discharge, and 3- the final disposal technologies:

The available volume reducing technologies include:

- Electrodialysis/Electrodialysis reversal
- Vibratory Shear - Enhanced Processing
- Precipitative Softening and Reverse Osmosis
- Enhanced Membrane System
- Brine Concentrator

Technologies which may be useful in this application but are still under development include:

- Two-pass Nanofiltration
- Forward Osmosis
- Membrane Distillation
- Capacitive Deionization

The zero liquid discharge technologies, on the other hand, include:

- Thermal processes
- Enhanced Membrane and Thermal processes
- Evaporation Ponds
- Wind-aided Intensified Evaporation

Final Disposal Options include:

- Disposal to Landfill
- Ocean Discharge
- Deep Well Injection
- Discharge to Waste Water Treatment Plant

Wiseman [29] underlined the criteria for evaluation of the desalination reject processing technology and the related pilot testing as follows:

1. Does the technology/pilot have regional applicability? Is the pilot implementable from regulatory, environmental, and funding perspectives?
2. Is the technology ready to be pilot tested?
3. Does the project have regional benefits?
4. How much water supply is saved by the project?

5. Does the project improve water quality or provide environmental benefits?
6. Can the technology be implemented for a full-scale project?
7. Are there barriers to full scale project implementation (regulatory, environmental, or funding?

3. Objectives, Aim and Scope of the Present Work

The main objective of the present research program is *the optimization of RO process efficiency and the decrease of consumption of the limited groundwater reserves through upgrading of the recovery of desalted water by adequate application of the most developed technologies of desalination membranes and chemicals.*

This chapter focuses on assessing the feasibility of increase of total RO recovery from the brine concentrate stream generated from RO by either secondary RO of the reject stream or by back recycling of reject to the initial RO feed, without significant sacrifice of permeate quality or excessive increase of product unit cost. Increase of total RO recovery means parallel decrease of the surface area of evaporation ponds required for disposal of the final reject stream which will enable a considerable saving in cost of plant installation.

In case of highly concentrated reject streams, the work includes an evaluation of the pretreatment processes required for attaining the highest possible recovery as e.g. removal or reduction of scale forming and gel forming ionic species and other membrane foulant components.

Other than promotion of process cost effectiveness, this investigation is also directed at promoting the environmental safety in relation to final reject disposal particularly in evaporation ponds, the commonly used approach for disposal of BWRO reject stream in Saudi Arabia. Reduction of the reject rate is expected to reduce the possibility of leak from these ponds and the pollution of groundwater reserves, also to control the frequent flooding of evaporation ponds to pollute the neighbouring habitat.

4. Experimental

Both laboratory and pilot NF testing were conducted.

The laboratory experimental system:

It consists of six test cells with circular turbulent agitation at the level of surface of membrane coupons installed in a test circuit which consisted of a low pressure pump, pressure gauge, cartridge filter, flowmeter and thermostated feed tank. Membrane samples were stored dry and thoroughly rinsed with deionized water before use. They were compacted in the distilled water at 120 psi, prior to testing, until steady flux is obtained, then conditioned by soaking in the testing solution for one hour. The testing feed pressures ranged from 80 to 100 psi. Tangential cross flow velocity ranged from 0.005 to 1 m/s and feed flux from 120 to 720 L/m^2.d.

The pilot testing unit:

Fig (1) shows schematic representation of the mobile pilot unit designed so as to enable conduction of NF and RO runs over a wide range of operation conditions, feed pressures, flow rates, pretreatment steps, and feasibility of reject recycling. Percent recovery was 85% except when otherwise stated. Both permeate and reject streams were recirculated back to the feed tank in order to keep steady feed water composition and concentration. Ionic concentrations were determined by ICP-AES (Parkin-Elmer, Boston, USA). Feed water temperature was thermostated at 25 C and pH was adjusted to the range 7.5 to 8.

Pilot site testing should enable:

- Direct connection to reject header or collection tank of existing desalination facilities for continuous treatment.
- Conduction of RO/NF pilot testing using:
- Conduction of desalination runs with different pretreatments for determination of the optimum recovery i.e. highest possible recovery attained under safe and steady operation performance.
- Optimization of operation conditions towards lower production cost, lower power and chemicals consumption.
- Investigation of reject treatment in different production sites in order to determine effect of reject characteristics and validity of selected technologies.

Chemical precipitation of radionuclides according to:

$$Ra^{2+}(trace) + BaCl_2 + SO^{2-}{}_4 = Ra \bullet Ba \bullet SO_4 + 2Cl^- \quad (1)$$

- Chemical precipitation of hardness components of reject stream by coagulation/settling.
- Conduction of NF runs for comparison of radionuclides and hardness rejection by NF to that obtained by chemical precipitation.
- Recycling of reject stream to the feed stream in the primary RO process.

5. General BW RO Reject Characteristics

- In addition to concentrated TDS, RO reject stream usually gets concentrated in hardness components and other polyvalent ionic species which are efficiently rejected in initial RO step as HMCs and radioactive isotopes.
- This stream is already sterilized and have passed already coarse and cartridge filtration.

- The unreacted pretreatment additives as antiscalants already concentrated in reject stream will lower the required dosing for scaling inhibition.
- pH and temperature values lie in the reasonable range for RO operation.
- Treatment of this stream either totally or partially would solve the problem of deficiency of evaporation ponds.
- Care should be taken for components which are harmful to RO process or membranes as Al, Fe, and Mn which become concentrated in the RO reject.

A typical reject streams analysis investigated in the present work is given by table (1)

Component	Concentration, mg/l	Parameter	Value
Ca^{2+}	2825.8	TDS	25,017.3 mg/l
Mg^{2+}	961.9	pH	7.6
Na^{+}	4406.3		
K^{+}	48.1		
NH_4^{+}	0		
NO_3^{-}	328.7		
Cl^{-}	10,030		
SO_4^{2-}	4837.2		
SiO_2	181.2		

Table 1. Typical Desalination Reject Water Analysis.

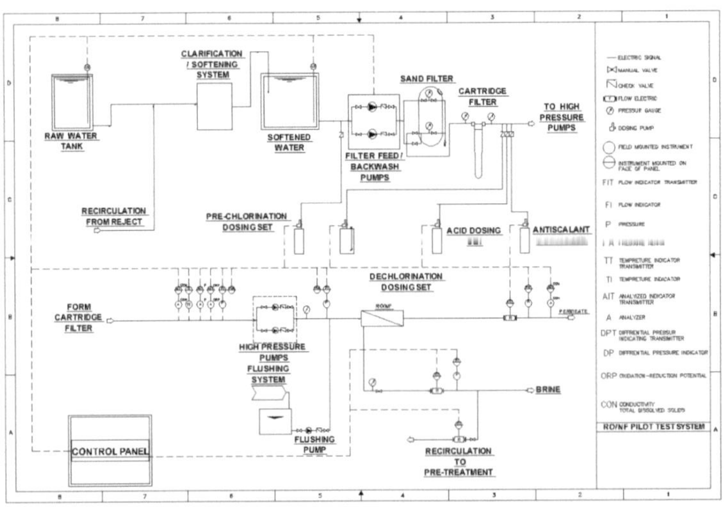

Figure 1. Schematic representation of the pilot testing unit.

5.1. Treatment of RO desalination reject stream by secondary RO or NF process

5.1.1. Process Definition:

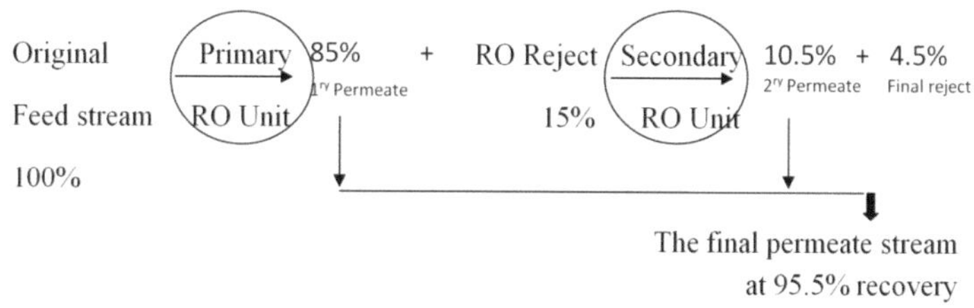

Figure 2.

According to Fig (2), if we consider the rate of feed stream to the initial RO treatment, e.g. raw well water, as 100% which is treated in the primary RO at percent recovery of e.g. 85%, the reject stream of 15% from the original feed will go for further processing in a secondary RO unit at a lower percent recovery of e.g. 70% the secondary permeate will be of 10.5% and the final reject will be reduced to 4.5% as referred to the original feed.

Upon blending of the primary + secondary RO permeate streams:

The total RO recovery becomes upgraded to much higher recovery (95.5% in the described case).

The final reject rate becomes reduced to less than the third of pervious reject rate and consequently the required surface area of evaporation pond.

The blending ratio is 8:1

The question is, how much higher is the cost per m³ of reject processing and what is its effect on the total process cost per m³ in view of the problems related to the treated reject i.e.:

1. Higher TDS.

2. The required higher feed pressure.

3. The possible higher cost of additives as specific Antiscalant.

In order to answer to those questions the various alternatives of reject treatment were investigated in detail.

5.1.2. Processing of Desalination Reject by Secondary RO:

Table (2) shows the results of RO treatment of three RO reject samples collected from different RO facilities of private sector and government water authorities in KSA of TDS of 33,370.4, 25,017.3 and 16,230.3 mg/l. Treatment is conducted by either brackish or sea water RO.

RO performance	Brine Concentration, mg/l					
	33,370.4		25,017.3		16,230.3	
	BWRO	SWRO	BWRO	SWRO	BWRO	SWRO
Initial Permeate TDS, mg/l	1025.6	292.7	581.3	204.4	485.5	131.2
Percent rejection	96.9	99.12	97.7	99.2	97.0	99.2
Feed Pressure, bar	35.18	43.59	35.3	48.78	29.36	37.09
Percent recovery of reject treatment	50	55	60	70	70	73
Total system recovery	92.5	93.25	94	95.5	95.5	95.95
Final permeate TDS, mg/l	187.0	125.0	122.7	87.3	86.2	52.9
Ratio of final reject to initial reject	0.50	0.45	0.40	0.30	0.30	0.27

Table 2. Results of RO treatment of three brine streams by BWRO & SWRO.

The higher the RO reject water TDS, the higher the required RO feed pressure particularly with SWRO. SWRO is shown to be the optimum selection in case of high salinity reject waters. It enables the highest recovery and lowest permeate TDS but required the highest operation pressure. SWRO is also useful in processing of reject water of high concentrations of undesired species as NO_3 or HMCs. Blending of the secondary RO permeate with the primary one is shown to realize the increase of total RO recovery with only a slight increase in final TDS in view of the low blending ratio. On the other hand, such reject processing in a secondary RO enabled the final reject rate to be remarkably reduced with consequent reduction of disposal cost.

Extent of RO reject processing and reduction of final brine rate is determined by the initial reject TDS and higher applied pressure and consequently the higher recovery realized upon use of sea water RO membrane elements.

5.1.3. Comparison of Performance of RO & NF in Processing of Desalination Reject Stream

For this comparative investigation pilot testing unified the main test conditions so that the different results reflect essentially the process behavior. A reject stream of 32,711 mg/l was treated by RO and NF systems having the same array adjusted to produce 1000 m^3/d, of course operated at different feed pressures, at the maximum attainable steady recovery. Final blending of the primary permeate (that of initial desalination unit) with the secondary permeate (that of the reject processing unit) was conducted to determine the total system recovery and the final product water quality. Comparison included also the extent of reduction of the final reject rate.

Parameters	**Sea Water RO**	**Brackish Water RO**	**Nanofiltration**
System Performance			
Salt rejection (%)	99.2	97.5	45.3
Permeate TDS, ppm	197.3	622.9	14,818
Fresh water Recovery (%)	71	63	80
% Rejection of some problem making components upon blending of primary and secondary permeate streams	Ca 99.73 NO_3 92.2 SiO_2 97.8	Ca 98.53 NO_3 85.31 SiO_2 97.85	Ca 89 NO_3 64.0 SiO_2 77.2
System cost factors			
Operation pressure, bar	50.87	37.33	14.94
Total system recovery (%)	95.65	94.45	97
Total Permeate TDS, ppm	88.58	129.8	1,899
Final reject rate, m^3/d	43.5	55.5	30

Table 3.

Results of Table (3) show that NF is operated at much higher recovery and much lower pressure than RO so that it could be operated by residual pressure of the reject stream. It is suitable, in fact, for intermediate reject treatment prior to a secondary RO desalination step or recycling in the feed of the primary RO unit. While NF has an only moderate TDS rejection, it rejects efficiently divalent or polyvalent species, organics and colloids [8]. A high hardness reject stream upon NF will, therefore, enable a subsequent RO treatment at a much higher percent recovery and lower operation pressure.

On the other hand, NF reject treatment upon blending would help to raise the primary RO permeate to a required TDS e.g. for drinking water level. The higher recoveries investigated with NF did not lead to higher TDS rejection.

5.1.4. Case Study of a 10,000 m^3/d BWRO Plant

In this plant the raw feed water have a radioactive contamination of 207.2 + 5.4 pCi/l of combined radium 226+228. It was requested *to increase the product rate to the maximum possible value by blending with conditioned feed stream while lowering the radioactivity to < 5 pCi/l the MCL of drinking water of the US- Environmental Protection Agency (EPA), with a final TDS higher than 300 ppm as a regional norm of drinking water TDS.* The present plant design failed to realize the required performance. On the other hand, the same site suffered flooding of evaporation ponds which was reported to be due to an over-estimated evaporation rate.

In fact, according to our previous results [6] the raw well water of TDS of 720.5 mg/l of this plant would be ideal for treatment by NF to produce the requested salinity since NF is characterized by an only modest rejection of TDS, but a rather high rejection of polyvalent ionic species as HMCs and radioactive isotopes [8]. However, in view of the important radioactive contamination, the concerned Water Authority selected RO, of much higher rejection than NF, to be conducted after a partial radionuclide separation by adsorption on the surface of hydrous manganese oxide (HMnO) according to:

$$2KMnO_4 + 3MnSO_4 + 2H_2O \rightarrow \underset{\downarrow}{5MnO_2} + 2H_2SO_4 + K_2SO_4$$

Results showed efficient rejection of both radionuclide and TDS to the level of drinking water, however, the value of product TDS was quite lower than 300 ppm.

In order to realize the required final product TDS increase the final product rate and simultaneously decrease the reject rate to the insufficient evaporation ponds partial treatment of the reject stream (already pressurized) by NF was investigated. Table (4) shows the resulting behavior.

	1	2	3	4	5	6(3+5)	7
Water stream	**Initial well water**	**After adsorption on HMnO**	**RO permeate**	**RO reject**	**Permeate of NF of reject**	**Final blend product rate**	**Final reject**
Rate m^3/d	15,552 to cooling towers	14,020 for both RO feed and blendinq streams	10,000	2,000	1,275 63,75% recovery	11,275	725
TDS ppm	720.5	720.5	70	4149	2508	406	
Ra $_{228+226}$ activity	207	82	1	547	26.8	4.08	

Table 4.

Results of pilot testing of reject treatment confirmed the realization of higher product rate at TDS > 300 mg/l and Ra activity less than the MCL.

5.2. Recycling of treated reject stream to the initial RO feed stream

For the case of already existing desalination facilities and the unavailability of space for additional reject processing unit, partial recycling of reject stream to the main feed stream aiming to upgrade the total recovery rate and reduce the final reject one is evaluated.

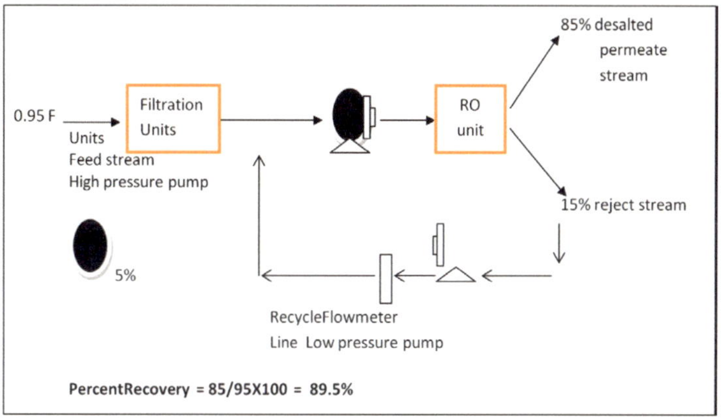

Figure 3.

The recycling circuit [9] Fig (3) consists of a low pressure pump, a control valve, and a flowmeter. It returns the required fraction of the reject stream ahead of the high pressure pump of the initial feed. The pilot plant was operated at various recycling rates. Upon recycling the reject, the total system working recovery remains at the previous value (85%) but from a higher feed TDS. A state of equilibrium is rapidly attained with a higher permeate TDS.

Water Components	RO feed	RO permeate	RO reject	RO feed 33.3% reject recycle	Secondary / permeate (33.3%)	RO feed 66.6% reject recycle	Secondary / permeate (66.6%)
Ca	284.08	3.2	1,882.22	365.16	4.25	460.83	5.73
Mg	95.85	1.2	641.09	121.72	1.81	154.33	2.3
Na	442.37	10.5	2,934.69	566.02	14.2	714.88	19.37
HCO_3	134.7	3.5	891.99	173.29	4.20	220.72	6.4
Cl	1010.2	20.1	6,692.61	1,292.95	27.08	1,635.92	36.02
SO_4	484.5	4.5	3,228.16	623.24	6.08	787.12	7.42
SiO_2	17.87	0.35	120.85	22.88	0.45	28.95	0.44
TDS	2,506.75	44.3	16,669.03	3,209.88	60.23	4,062.30	76.8

Table 5. Variation of secondary feed & permeate TDS with percent recycle.

However, in order to make the calculated percent recovery expressive of the saving in feed water from the wells and of the decrease in final reject stream i.e. representative of the promoted process efficiency, we adopted [9] referring the permeate rate to the lowered raw water feed rate in calculation of recovery.

Table (5) describes a pilot test of BWRO dealing with a groundwater of 2,520.0 mg/l which results in a permeate water of 82.5 mg/l and reject water of 16,790.0 mg/l at 85% recovery. The first three columns give the analysis of each of these streams. Column no. 4 shows the analysis of the increased RO feed TDS upon recycling of 33.3% of the reject stream to initial RO feed one. The corresponding permeate analysis is given by column no. 5 column no. 6 and 7 give the corresponding results for the recycling of 66.6% of the reject stream to the RO feed.

Results revealed than *partial recycling of the reject stream introduced only a moderate increase of the individual ion concentrations in RO feed stream (despite the high reject TDS) in view of the dilution of the recycled fraction of reject upon mixing with the whole feed stream.* In already existing BWRO facilities, therefore, partial reject recycling is shown to raise the percent recovery, lower the consumption of raw feed water, and to lower remarkably the final reject rate and consequently the required land area and cost of installation of evaporation ponds without significant sacrifice of product water quality.

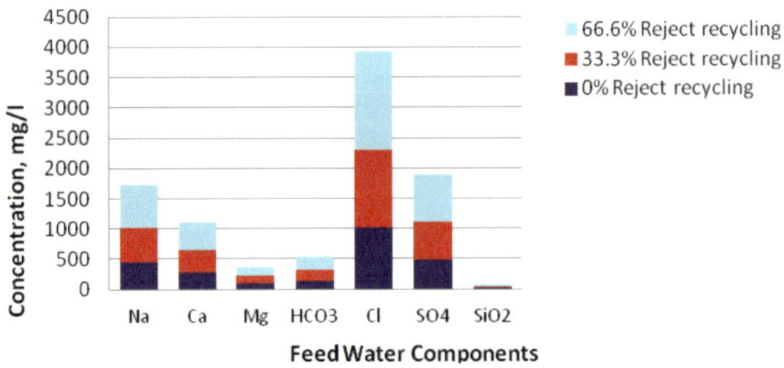

Figure 4. Variation of concentration of component species of the RO feed with percent reject recject recycling.

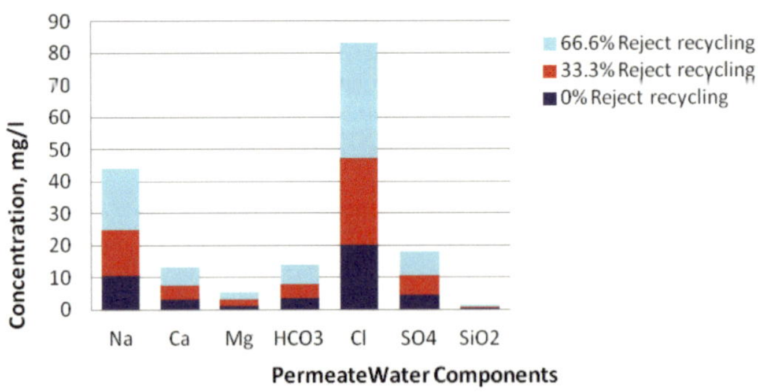

Figure 5. Variation of concentration of component species of the RO permeate with percent reject recycling.

Fig (4) shows the variation of concentration of the RO feed component species with percent reject recycling. These values correspond to an increase of feed TDS from 2,506.8 mg/l to 3,209.9 mg/l by recycling of 33.3% of the reject stream, then to 4,062.3 ppm by increase of recycling to 66.6%. Fig (5), on the other hand, shows the corresponding variation of the concentration of the water species in the permeate stream upon recycling of reject at the mentioned rates. According to these results the increase of permeate TDS parallel to increase of feed TDS upon recycling of reject to original feed stream is limited and did not compromise the drinking water quality. The recycling of 66.6% of the reject raised the permeate TDS only from 44.3 to 76.8 ppm.

As for antiscalant dosing during RO reject processing, in principle the antiscalant which is concentrated in the reject stream is useful for the subsequent reject processing. However, with the higher concentration of certain scale forming components like SiO_2 in the reject, a different type of antiscalant became required to cover the saturation during the reject processing.

As an example, the general validity antiscalant (Genesys LF) was used in the primary BWRO step of the raw well water of 2,506.8 mg/l (1,000 m^3/d) operated at 85% recovery at a dose of 3.03 mg/l. for the reject processing, on the other hand, (150 m^3/d) of a TDS of 16,230.3 mg/l and at higher concentration of different components particularly SiO_2, a SiO_2 specific antiscalant was required at a rate of 11.42 mg/l consideration. The difference in price between the different dosed antiscalants did not add much to the general cost/m^3 (<1% increase).

5.3. Desalination Reject Processing by Chemical Softening Prior to Recycling or Secondary RO Treatment

After RO of high salinity groundwaters, processing of reject stream by chemical softening or NF aims to remove or reduce the scale forming components accumulated during RO so as to enable the promotion of the total process recovery through subsequent secondary RO step or partial recycling of treated reject to the initial RO feed.

Reject water rather high in Ca, Mg and SiO_2 can be softened by addition of hydrated lime, $Ca(OH)_2$ and sodium carbonate which settles out of water $CaCO_3$ and after all of HCO_3^- is consumed, the remaining OH- combines with Mg^{2+} to deposit $Mg(OH)_2$ on which surface SiO_2 is removed as an adsorption complex (10). Results have shown that for high SiO_2 reject streams additional Mg may have to be added in order to attain the required SiO_2 removal.

5.3.1. Partial Cold Lime Softening (CLS)

Fig (6) shows typical results of partial CLS which consists in dosing only hydrated lime to the RO reject water. For each species the first column to the left represents the concentration in the reject water and the second shows the effect of dosing of $Ca(OH)_2$, concentrations are represented as ppm $CaCO_3$. When reject pH was raised from 8.3 to 10.0 by lime dosing, the precipitation which took place resulted in reduction of Ca^{2+} content by 56.5%, M alkalinity by 70% the remainder being as CO_3^{2-}, and complete consumption of HCO_3^-. On the other hand, P alkalinity increased by 140 ppm while other reject water components including Mg and SiO_2 remained unchanged. TDS was lowered by 26.35% depending on extent of conducted lime dosing.

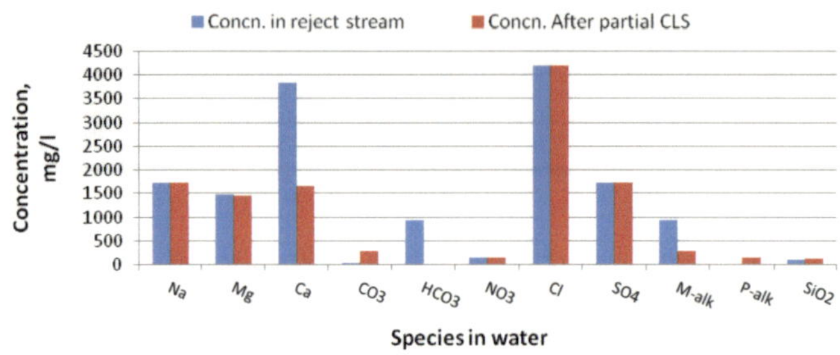

Figure 6. Influence of partial CLS on RO reject composition.

According to these results the advantages of the partial CLS are:

1. Ease of operation with only one dosing and coagulation step.
2. Lower cost of chemical dosing than lime-soda ash CLS.
3. Parallel lowering of TDS by precipitation lowers the desalination load on the subsequent reject processing.
4. It is particularly interesting in case of reject streams where Mg and consequently SiO_2 removal do not represent a problem for processing.

5.3.2. *Conventional Cold Lime Softening*

Fig (7), on the other hand, shows the effect of addition of Na_2CO_3 after the initial partial CLS. For each species the first column to the left represents the concentration in the RO reject water, the second shows the effect of dosing of Na_2CO_3 at a concentration of 45% of the lime concentration previously added during the partial CLS. This lowered Ca concentration by 75.3%. Mg was practically not removed at this level of alkalinity in view of the absence of additional free OH^- for deposition of $Mg(OH)_2$.

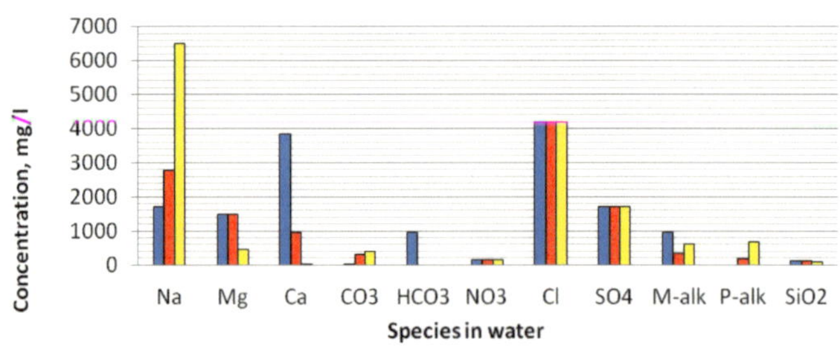

Figure 7. Influence of Conventional CLS with different closing rates of CaCO3 on RO reject composition.

The third bar of Fig (7) belongs to dosing of an excess of Na_2CO_3 to attain the double of concentration of lime of the initial partial CLS in order to raise alkalinity to a quite higher level.

http://dx.doi.org/10.5772/50234

Such increase of alkalinity did not lead to any further deposition of Ca. *In fact, our CLS results showed a minimum Ca concentration (22 ppm) at which higher lime-soda ash dosing had no effect.* Fig (7) shows in parallel a considerable increase of Na, a lower increase of CO_3 and a decrease of Mg of 67% in view of the additional free OH^-. SiO_2 is lowered by 22% by adsorption on the deposited $Mg(OH)_2$. Complete CLS resulted in decrease of reject water TDS by 7.6% with respect to original reject water TDS.

It is worthy to notice that stoichiometrically equivalent concentrations of coions $CO_3^{\ 2-}$ and OH^- to those of Ca^{2+} and Mg^{2+}, or higher, are required for precipitation of $CaCO_3$ and $Mg(HO)_2$. As precipitation advances, however, alkalinity as well as supersaturation are reduced. In order to achieve a steady rate of precipitation and residence time typically between 60 to 90 minutes, we had to keep a supersaturation factor (SSF) of at least three.

Parallel to chemical softening and in the same reactor, components like HMCs which may be concentrated in RO reject were shown to be better precipitated through dosing of sulphide since their sulphides are more insoluble than their hydroxides or carbonates, similarly, chlorine (hypochlorite), added during softening improved removal of Fe^{2+} by oxidation to the Fe^{3+}, Or sulphite improved precipitation of the soluble Cr^{6+} by reduction to Cr^{3+}.

5.4. Reject Processing after CLS or NF

After each of partial or conventional CLS or NF of the RO reject stream, further processing was conducted through either partial recycling to the feed stream of the primary RO unit or feeding an independent secondary RO unit.

Fig (8) shows the results of recycling of softened reject stream (partial CLS) in the range of 25 to 75 percent to the feed stream of the primary RO unit. Recycling increased feed TDS which was shown to have only limited influence on permeate TDS. *While at 75% recycling the feed TDS was practically doubled to attain 5461.7 mg/l, treated under mainly similar conditions by the same pilot RO unit operated using High rejection, low energy RO membranes, the permeate TDS showed an only limited increase from 60.9 to 139.3 mg/l which does not compromise its quality for subsequent application.*

According to these results, already present RO facilities, without need of additional equipment or space, can promote the total system recovery and reduce the final reject rate and consequently the cost of the waste disposal through a simple system modification without significant sacrifice of RO permeate quality.

On the other hand, if reject processing aims to increase the final product rate, the softened stream can be treated in an independent secondary RO unit. *The comparison between RO performance of the softened and the unsoftened reject streams shows that presoftening is particularly interesting in case of high TDS, high hardness brines.* Table (6) for an RO reject of 25,017.3 mg/l having a total hardness of 11,007.0 mg/l as $CaCO_3$ required a much smaller RO system array to result in a lower TDS permeate at a much higher recovery than the same reject stream after softening (TDS = 18,435 mg/l, total hardness = 8,279.3 mg/l as $CaCO_3$).

RO Performance	Unsoftened Reject Stream	Softened reject Stream
RO System Array*	(13:6:3)6	(10:5:2)6
Percent Recovery	53%	62%
Permeate TDS, mg/l	763.4	552.3

*pressure vessels of 6 RO elements each arrayed in three stages.

Table 6. Comparison of RO results of presoftened and unsoftened reject.

On the other hand, for NF of reject prior to secondary RO treatment and the efficient dehardening by NF in addition to partial TDS rejection, it enabled recoveries as high as 85% of the secondary RO. This resulted in total process recovery as high as 97.75%.

Results showed that treatments of RO reject by NF prior to recycling or treatment in secondary RO unit is particularly interesting in case of medium salinity and total hardness reject streams while for highly concentrated reject streams CLS is more effective and has a lower cost than NF.

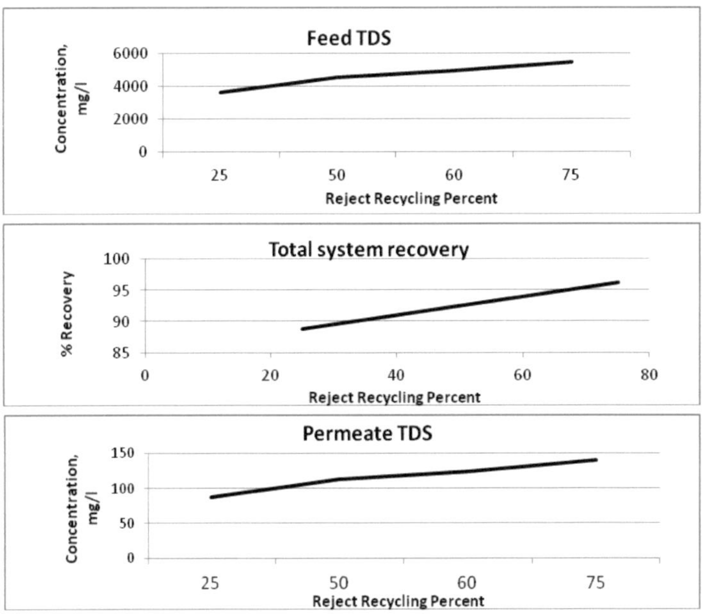

Figure 8. Effect of Partial Recycling of Softened Reject Stream.

5.5. Comparison between Removal of Scale Forming Components from RO reject by NF and by CLS

Removal of hardness components concentrated in RO reject as Ca, Mg, SO_4 together with SiO_2, Fe and Mn as well as other possible components like HMC's, was investigated by NF in comparison with precipitation by the conventional CLS.

In order to conduct the comparison of the two methods under similar conditions the extent of rejection recorded by NF was the basis of selection of the dosing rate of lime and soda ash which realize the same Ca rejection. In fig (9) for each species, the first column to the left represent the initial concentration in the reject stream, the second and the third represent the results of rejection by NF, and softening, respectively.

While NF lowered the concentration of all the species to various extents and consequently lowered TDS, softening lowered only that of components included in the softening reactions as HCO^{3-}, P-alkalinity and SiO_2 [11]. Softening raised, on the other hand, concentration of Na+, $CO_3{}^{2-}$, and M-alkalinity. As for SiO_2, which is directly rejected by NF, it is removed upon softening by adsorption on $Mg(OH)_2$ deposited surface at high pH values, but at a lower efficiency than NF rejection.

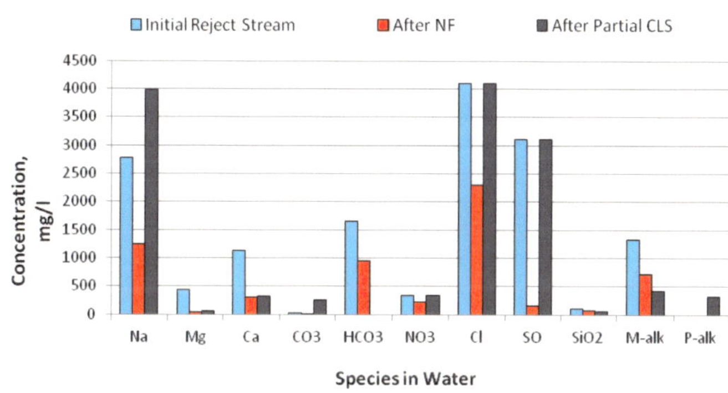

Figure 9. Comparision between NF & partial CLS in processing of reject stream.

While the chemical softening is usually stated as having lower cost [10], [12] the detailed consideration of all the related cost factors or additional process steps that are not included in NF and which should be added to the cost of softening in order to realize the same performance as NF, revealed the cost advantage of NF reject treatment. Chemical softening but not NF requires stoichiometric or higher dosage of lime and soda ash to reduce hardness, disposal of large amounts of sludge which may include dosage of polyelectrolytes and/or sludge conditioning before delivery to settling ponds and landfill disposal, raising of pH of the reject stream up to > 9.5 for indirect removal of SiO2 after deposition of $Mg(OH)_2$ and sophisticated installations for chemical dosage, and settling tanks. Our results have shown that CLS is not as complete as by IER or NF for removal of Ca and does not effectively remove organics, or reduce TDS.

The above considerations extend to the treatment of different types of industrial WW's which contain hardness components, HMC's, may be together with organics and suspended solids where NF application will be optimum particularly if complete desalination is not required.

6. Conclusions

- Processing of the desalination reject stream, instead of just getting rid of it, is conducted by laboratory and pilot testing in order to promote the desalted water recovery and reduce the final reject disposal problems and costs which will increase the total desalination process efficiency, cost effectiveness and environmental safety.

- Among the investigated processing alternatives the most efficient ones in case of medium concentration brine stream (up to 10,000 mg/l) are (a) (high rejection low energy RO + use of specific antiscalant), (b) partial recycling of reject to the feed stream of the initial RO unit.

  In already present RO facilities, reject recycling does not require extra footprint. Results showed that percent reject recycling as high as 75% did not significantly increase the final permeate salinity. For new projects, on the other hand, increase of total product rate was realized through a secondary RO treatment of reject.

- In case of high TDS reject streams up to 33,000 mg/l, reject processing by partial cold lime method, conventional cold lime method or nanofiltration was conducted prior to circulation to initial RO feed or treatment in secondary RO unit. Results confirmed the promotion of total percent recovery without significant sacrifices of total permeate qualities.

- Partial CLS is particularly interesting in case of reject streams where Mg and SiO_2 removal do not represent a problem for processing. Beside ease of operation and lower cost than conventional CLS, a partial CLS lead to higher decrease of reject TDS and does not increase Na concentration.

Author details

M. Gamal Khedr

Address all correspondence to: mgakhedr@yahoo.fr

National Research Centre, Cairo, Egypt

References

[1] Khedr, M. Gamal. (2000). Membrane Fouling Problems In Reverse Osmosis Desalination Applications. *Desalination &Water Reuse*, 10(3), 9.

[2] Khedr, M. Gamal. (1978). The Rejection of Scale Forming Ions From Water Containing Salt Mixtures By Reverse Osmosis. *Chemie-Ingenieur-Technik*, 51, 516.

[3] Khedr, M. Gamal. (2004). Optimization of Reverse Osmosis Process Efficiency and Environmental Safety through Reject Processing. Hamburg. *Euromenbrane International Conference*, 600.

[4] Khedr, M. Gamal. (2009). Desalination and Water Treatment, , 2, 342.

[5] Der Bruggen, B. V., Vandecasteele, C., Gestel, T. V., Doyen, W., & Ley san, R. (2003). Environmental Progress, , 22(2), 6.

[6] Richards, A., Suratt, W., Winters, H., & Kree, D. (2001). J. AWWA, 01.

[7] Khedr, M. Gamal. (2011). Processing of Desalination Brine by Single or Hybrid Membrane Processes for Optimization of Process Efficiency, Cost Effectiveness and Environmental Safety. First International Conference on Desalination and Environment: A Water Summit, Elsevier, Abu Dhabi Oct (2011)

[8] Ahmed, M. W. H., Shayya, D., Hoey, D., Maendran, A., Morris, R., & Al-Handaly, J. (2000). *Desalination*, 130, 155.

[9] Al-Akbany, F., Al-Mutairi, F., & Al-Jamaan, A. (2005). Study of Reuse RO Rejected Water. *Water and Sewage Authority of Riyadh.*

[10] Sommariva, C., Frederica, A., Mosto, N., & Mac Donald, Mott. (2009). UAE, European Desalination Society. *Euromed Conf., D and WR*, 18(4).

[11] Smith, D., & Humphrey, S. (2001). *CSIRO Land and Water, Griffith NSW (2000), Research Rpt.*

[12] Ahmed, M., et al. (2003). *Desalination*, 158, 109.

[13] Shahalam, A. (2009). *European Journal of Scientific Research*, 28(4), 514.

[14] Heijman, S. G. J., Guo, H., Li, S., Van Dijk, J. C., & Wssels, L. P. (2009). *Desalination*, 236, 375.

[15] Jeppesen, T., Shu, L., Keir, G., & Jegathee, S. (2009). *Journal of Cleaner Production*, 17(7), 703.

[16] Ersever, I., & Pirbazari, M. M. (2002). *California Energy Commission*, www.energy.ca.gov/reports/2004-04-02.

[17] Queen, A., Robinson, J., & Haas, W. (2004). *Patent application N◦ 10840249 filed on 05/07/2004, US classes, 210/259, GE Global patent operation.*

[18] Summit Industries, Oil and Gas Industry Service Company. (2006). USA.

[19] Redetzke, D. J. (2002). *Independent Salt Company*, Kansas, USA.

[20] Corral, A. F., & Yenal, U. (2009). *Minimization of RO reject stream through VSEP, Vibratory Shear enhanced process*, www.pdfio.com/k-426648html.

[21] Arnold, G. R. (2008). *Desalination of Central Arizona Project Water, so of Vibratory Separation Enhanced Process (VSEP), for Water Recovery from RO brine; Maximization of Water*

Recovery using a combination of processes, Technology and Research Initiative Fund (TRIF).

[22] Cates, T., Dickie, B., & Bang, M. (2007-2008). Dept. of Chem. Engineering, Univ. Of Saskatchewan, *Final Report.*

[23] Priya, M. N., & Palanivelu, . (2006). *Indian Journal of Chemical Technology,* 13, 262.

[24] European Union Reclaim Water Project. (2010). Treatment and Recovery of RO Brine for higher Recovery in NEWater Factories. *the EU's sixth Framework Programme for Research and Technological Development,* Singapore, PUB.

[25] Lee, I. Y., Ong, S. L., Tao, G., Viazanath, Viawanath B., Kekre, K., & Seah, H. (2008). *Water Science and Technology,* 58(4), 931.

[26] Lee, I. Y., Ong, S. L., Tao, G., Kekre, K., Viswanath, B., Lay, W., & Seah, H. (2009). *Water Research,* 43(16), 3948.

[27] Duraflow Co. (2008). RO Brine Recovery; A California Power-Plant Converts to Duraflow membranes for Water Recycling System. *Application Bulletin.*

[28] Kepke, J. T., Foster, L., Cesca, J., & Mc Cann, D. (2007). Australia. *Second International Salinity Forum.*

[29] Weiseman, R. (2010). *Desalination and Water Reuse,* 20(1), 14.

Chapter 7

Nanofiltration Process Efficiency in Liquid Dyes Desalination

Petr Mikulášek and Jiří Cuhorka

Additional information is available at the end of the chapter

http://dx.doi.org/10.5772/50438

1. Introduction

Membrane science and technology has led to significant innovation in both processes and products over the last few decades, offering interesting opportunities in the design, rationalization, and optimization of innovative production processes. The most interesting development for industrial membrane technology depends on the capability to integrate various membrane operations in the same industrial cycle, with overall important benefits in product quality, plant compactness, environmental impact, and energetic aspects.

The membrane separation process known as nanofiltration is essentially a liquid phase one, because it separates a range of inorganic and organic substances from solution in a liquid – mainly, but by no means entirely, water. This is done by diffusion through a membrane, under pressure differentials that are considerable less than those for reverse osmosis, but still significantly greater than those for ultrafiltration. It was the development of a thin film composite membrane that gave the real impetus to nanofiltration as a recognised process, and its remarkable growth since then is largely because of its unique ability to separate and fractionate ionic and relatively low molecular weight organic species.

There are probably as many different applications in the whole chemical sector (including petrochemicals and pharmaceuticals) as in the rest of industry put together. Many more are still at the conceptual stage than are in plant use, but NF is a valuable contributor to the totality of the chemicals industry. The production of salt from natural brines uses NF as a purification process, while most chemical processes produce quite vicious wastes, from which valuable chemicals can usually be recovered by processes including NF. The high value of many of the products of the pharmaceutical and biotechnical sectors allows the use of NF in their purification processes [1,2].

Reactive dye is a class of highly coloured organic substances, primarily used for tinting textiles. The dyes contain a reactive group, either a haloheterocycle or an activated double bond, which, when applied to a fibre in an alkaline dye bath, forms a chemical bond between the molecule of dye and that of the fibre. The reactive dye therefore becomes a part of the fibre and is much less likely to be removed by washing than other dyestuffs that adhere through adsorption. Reactive dyeing, the most important method for the coloration of cellulosic fibres, currently represents about 20-30% of the total market share for dyes, because they are mainly used to dye cotton which accounts for about half of the world's fibre consumption.

Generally, reactive dyes are produced by chemical synthesis. Salt, small molecular weight intermediates and residual compounds are produced in the synthesis process. These salt and residual impurities must be removed before the reactive dyes are dried for sale as powder to meet product quality requirement. Conventionally, the reactive dye is precipitated from an aqueous solution using salt. The slurry is passed through a filter press, and the reactive dye is retained by a filter press. The purity of the final reactive dye product in conventional process is low, having a salt content around 30%. Furthermore, the conventional process is carried out in various batches, which makes the entire process highly labor intensive and causes inconsistency in the production quality.

In dye manufacture, like most other processes, there is a continual search for production methods that will improve product yield and reduce manufacturing costs. Dye desalting and purification, the process by which impurities are removed to improve the quality of the product, is currently one of the biggest applications for NF technology. Dye manufacturers are now actively pursuing the desalting of the finished dye prior to spray drying because it not only improves product quality, but makes spray drying more efficient because the granulation of the dye takes place without the production dust. NF is proving to be an ideal method for this salt removal [3,4].

Nanofiltration is the most recently developed pressure-driven membrane separation process and has properties that lie between those of ultrafiltration (UF) and reverse osmosis (RO). The nominal molecular weight cut-off (MWCO) of NF membranes is in the range 200-1000 Da. Separation may be due to solution diffusion, sieving effects, Donnan and dielectric effects. The rejection is low for salts with mono-valent anion and non-ionized organics with a molecular weight below 150 Da, but is high for salts with di- and multi-valent anions and organics with a molecular weight above 300 Da. Thus, NF can be used for the simultaneous removal of sodium chloride (salt) and the concentration of aqueous dye solutions [5,6].

Diafiltration is the process of washing dissolved species through the membrane, which is to improve the recovery of the material in permeate, or to enhance the purity of the retained stream. Typical applications can be found in the recovery of biochemical products from their fermentation broths. Furthermore, diafiltration can be found in removal of free hydrogel present in external solution to purification of a semi-solid liposome (SSL), purification of polymer nanoparticles, enhancing the protein lactose ratio in whey protein products, separating sugars or dyes from NaCl solution (desalting), and many other fields. According to the property of the solute and the selectivity of membrane, diafiltration can be used in the process of MF, UF or NF [7-12].

http://dx.doi.org/10.5772/50438

The aim of this study is also devoted to the mathematical modelling of nanofiltration and description of discontinuous diafiltration by periodically adding solvent at constant pressure difference.

The proposed mathematical model connects together the design equations and model of permeation through the membrane. The transport through the membrane depends on the different approaches. Firstly the membrane is regarded to a dense layer and in this case transport is based on solution-diffusion model [13,14]. Second approach is regarded membrane to porous medium. Models with this approach are based mainly on extended Nernst-Planck equation. Through this approach, a system containing any number of n ions can be described using set of (3n + 2) equations. In this approach, it is assumed that the flux of every ion through the membrane is induced by pressure, concentration and electrical potentials. These models describe the transport of ions in terms of an effective pore radius r_p (m), an effective membrane thickness/porosity ratio $\Delta x/A_k$ (m) and an effective membrane charge density X_d (mol/m^{-3}). Such a model requires many experiments for determination of these structural parameters. These models are hard to solve [6,7,12,15]. The last approach is based on irreversible thermodynamics. These models assume the membrane as "black box" and have been applied in predicting transport through NF membranes for binary systems (Kedem-Katchalsky, Spiegler-Kedem models). Perry and Linder extended the Spiegler and Kedem model to describe the salt rejection in the presence of organic ion. This model describes transport of ion through membrane in terms of salt permeability P_s, reflection coefficient σ [10,12,16-18]. In our work is solution-diffusion model used. The solution-diffusion model can be replaced by more theoretical model in future.

1.1. Theoretical model

Salt rejection of a single electrolyte has been described by Spiegler and Kedem [19] by the three transport coefficients: water permeability L_P, salt permeability P_S and reflection coefficient σ. For the curve describing salt rejection as a function of flow, the salt can be treated as a single electroneutral species.

Assuming linear local equations for volume and salt flows, these authors derived an expression of salt rejection R_S as a function of volume flux J_V. The local flux equations are:

$$J_V = -L_P\left(\frac{dp}{dx} - \sigma\frac{d\pi}{dx}\right) \tag{1}$$

$$J_S = -P\frac{dc_S}{dx} + (1-\sigma)c_S J_V \tag{2}$$

where salt rejection R_S is defined by the salt concentrations c_F and c_P in the feed and permeate streams respectively.

$$R_S = 1 - \frac{c_P}{c_F} \tag{3}$$

and

$$c_P = \frac{J_S}{J_V} \tag{4}$$

With constant fluxes, constant coefficients P and σ and with condition in Eq. (4), integration of Eq. (2) through the membrane thickness yields:

$$\frac{J_V(1-\sigma)\Delta x}{P} = \ln\frac{c_P\sigma}{c_P - c_F(1-\sigma)} \tag{5}$$

and then salt rejection can be expressed as:

$$R_S = \frac{(1-F)\sigma}{1-\sigma F} \tag{6}$$

where

$$F = e^{-J_V A},\ A = \frac{1-\sigma}{P}\Delta x = \frac{1-\sigma}{P_S} \tag{7}$$

where P is the local salt permeability and P_S is overall salt permeability.

In a mixture of electrolytes, the interactions between different ions can be very important and the behaviour of mixed solutions in NF cannot be predicted from the coefficients describing each salt separately. The differences in the permeabilities of ions lead to an electric field, which influences the velocity of each ion. Thus one needs to analyse all ion fluxes together. This analysis for the mixture of two electrolytes with common permeable counter ion and two co-ions, which are the rejected and the permeable ion respectively, is presented bellow.

1.2. Salt permeation in presence of retained organic ions

Consider a system (see Figure 1) which consists of semipermeable membrane separating two aqueous solutions with mixed electrolyte sharing a common permeable cation (*1*) and two anions which anion (*2*) is permeable through the membrane and anion (*3*) is fully rejected.

For the sake of simplicity let us consider a mixture of a mono-monovalent salt (*NaCl*) and a multifunctional organic anion C_x^{-v} containing v negatively charged groups per molecule in a sodium salt form.

http://dx.doi.org/10.5772/50438

The two electrolytes are fully dissociated as shown in Eq. (8).

$$\begin{aligned} NaCl &\Leftrightarrow Na^{+} + Cl^{-} \\ C_x Na_v &\Leftrightarrow vNa^{+} + C_x^{-v} \end{aligned} \tag{8}$$

If we assume the salt as a single electroneutral species we can express the condition of equilibrium between feed and permeate solution as:

$$(a_1 a_2)_F = (a_1 a_2)_P \tag{9}$$

We can also consider the equilibrium between feed solution and the solution inside the membrane. Then the equilibrium condition can be expressed as:

$$(a_1 a_2)_F = (a_1 a_2)_M \tag{10}$$

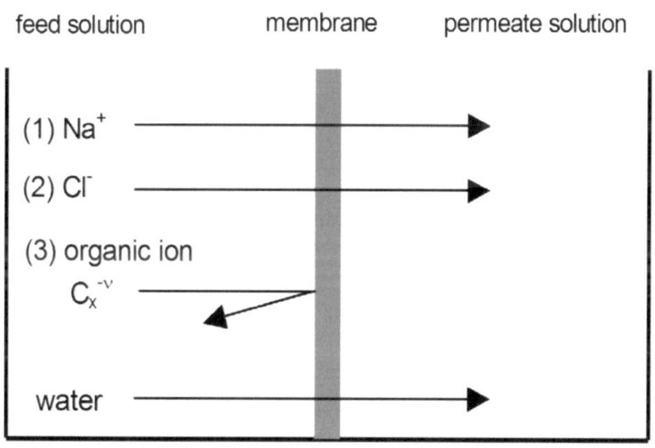

Figure 1. Scheme of the system.

In the feed solution containing a nonpermeable multifunctional organic anion at a concentration C_x, the following conditions of electroneutrality can be written for each phase:

$$[Cl\ ]_F = c_{SF} \tag{11}$$

$$[X^{-}]_F = C_{xF} \tag{12}$$

$$[Na^{+}]_F = c_{SF} + vC_{xF} \tag{13}$$

$$[Cl^{-}]_M = [Na^{+}]_M = c_{SM} \tag{14}$$

We assume equilibrium on the membrane solution interface, approximating activities with concentrations and substituting Eqs. (11) and (14) into Eq. (10), one obtains the value of salt concentration inside the membrane on the feed side:

$$c_{SM} = c_{SF}\left(1 + \frac{\nu C_{xF}}{c_{SF}}\right)^{0.5} \tag{15}$$

The expression for c_{SM} from Eq. (15) is now used to integrate Eq. (5). In other words, in the presence of Donnan exclusion forces, induced by the impermeable organic ions, the salt transport across the membrane proceeds as if the membrane was exposed to a salt solution having a concentration c_{SM} instead of c_{SF}. Thus the value of c_{SM} and not of c_{SF} determines the driving force for the salt passage and should be used as boundary condition during the integration of Eq. (1) and Eq. (2).

Then the expression for salt rejection in the presence of retained organic ion can be written as:

$$R_S = \frac{(1-\sigma F) - (1-\sigma)\left(1 + \frac{\nu C_{xF}}{c_{SF}}\right)^{0.5}}{1-\sigma F} \tag{16}$$

1.3. Concentration dependence of the solute permeability

The concentration dependence of the solute permeability was proposed by Schirg and Widmer [16] as an exponential function:

$$P = \alpha c_F{}^{\beta} \tag{17}$$

where c_F is the concentration of the permeable component in the feed [g.l^{-1}],

α - coefficient for salt permeability [m.s^{-1}],

β - coefficient for concentration dependence of salt permeability [-].

With introducing of Eq. (17) into Eq. (6), the retention for single electrolyte can be written as:

$$R = 1 - \frac{1-\sigma}{1 - \sigma \exp\left(\frac{(\sigma - 1)J_V}{\alpha c_F{}^{\beta}}\right)} \tag{18}$$

Similar, with introduction of Eq. (17) into Eq. (16), the salt retention for the system with retained organic ion can be expressed as:

http://dx.doi.org/10.5772/50438

$$R_S = 1 - \frac{(1-\sigma_S)\left(1+\frac{\nu M_S C_{xF}}{M_x c_{SF}}\right)^{0.5}}{1-\sigma_S \exp\left(\frac{(\sigma_S-1)J_V}{\alpha_S c_{SF}{}^{\beta_S}}\right)} \tag{19}$$

1.4. Mathematical modelling of diafiltration

Mathematical model connects together balance equations and solution-diffusion model, which is extended by dependence of salt permeability on the salt concentration in feed and Donnan equilibrium.

The balances for the concentration mode can be written as:

Solvent mass balance:

$$\frac{d(V_F \rho_F)}{d\tau} = -J A^* \rho_P \tag{20}$$

Mass balances of dye and salt:

$$\frac{d(V_F c_{D,F})}{d\tau} = -J A^* c_{D,P} \tag{21}$$

$$\frac{d(V_F c_{S,F})}{d\tau} = -J A^* c_{S,P} \tag{22}$$

Eq. (20) is possible to write in the form:

$$\frac{dV_F}{d\tau} = -J_V A^* \tag{23}$$

Mass balances of dye and salt are formally same and we can solve them together. Subscripts i represent dye and salt. Eq. (21) (or (22)), may be re-written as:

$$\frac{d(V_F c_{i,F})}{d\tau} = -J_V A^*(1-R_i)c_{i,F} \tag{24}$$

where R_i is real rejection.

In the concentration mode, the volume and the concentration in feed depends on the time. Expanded differential equation with using the product rule can be written as:

$$V_F \frac{dc_{i,F}}{d\tau} + c_{i,F} \frac{dV_F}{d\tau} = -J_V A^*(1 - R_i)c_{i,F} \tag{25}$$

Substituting Eq. (23) into Eq. (25) leads to:

$$V_F \frac{dc_{i,F}}{d\tau} = J_V A^* R_i c_{i,F} \tag{26}$$

Dividing Eq. (26) by Eq. (23) leads to:

$$c_{i,F} = c_{i,F0} \left(\frac{V_{F0}}{V_F} \right)^{R_i} \tag{27}$$

If we assume constant rejection and permeate flux (for small change of volume in feed tank, or better of yield - permeate volume divided by feed volume, it is achieved) or the average values integrations of Eq. (27) and Eq. (23) with the boundary conditions (V_{F0} to V_F) resulting in Eq. (28) and Eq. (29):

$$c_{i,F} = c_{i,F0} \left(\frac{V_{F0}}{V_F} \right)^{R_i} \tag{28}$$

$$\tau = \frac{V_{R0} - V_R}{J_V A^*} \tag{29}$$

On the base of Eq. (28) and Eq. (29) we can obtain the concentration in feed tank and the time for separation of certain permeate volume in concentration mode, respectively. Next process is diluting. Pure solvent (water) is used as diluant. Salt concentration in feed tank after this operation (c_S') is:

$$c_S' = c_{i,F0} \left(\frac{V_F}{V_{F0}} \right) \tag{30}$$

This concentration (c_S') is now equal to the salt concentration in feed tank ($c_{S,F0}$) for the next concentration mode in the second diafiltration step.

For solving of these equations we need to know dependence of rejection and permeate flux on salt concentration in feed.

The basic equations for rejection can be written as:

$$J_S = B(c_F - c_P) \tag{31}$$

$$c_P = \frac{J_S}{J_V} \tag{32}$$

This model can be extended by the dependence of salt permeability on salt concentration in the feed [17]. To avoid some in conveniences with units, here c* is introduced and chosen to be 1 g/l.

$$B = \alpha \left(\frac{c_{S,F}}{c^*} \right)^{\beta} \tag{33}$$

Assuming equilibrium on the membrane - solution interface we can obtain (approximating activities with concentrations) [17]:

$$c_{S,W} = c_{S,F} \left(1 + \frac{v_D . c_{D,F} . M_S}{c_{S,F} . M_D} \right) \tag{34}$$

In the presence of Donnan exclusion forces, induced by the impermeable organic ions, the salt transport across the membrane proceeds as if the membrane were exposed to a salt solution having concentration $c_{S,W}$ instead $c_{S,F}$. Thus the value of $c_{S,W}$ and not that of $c_{S,F}$ determines the driving force for the salt passage.

Then the expression for salt passage in the presence of retained organic ion can be written as:

$$J_S = \alpha \left(\frac{c_{S,F}}{c^*} \right)^{\beta} (c_{S,W} - c_{S,P}) \tag{35}$$

and then salt concentration in permeate can be expressed as

$$c_{S,P} = \frac{\alpha \frac{c_{S,F}^{\beta+1}}{c^{*\beta}} \left(1 + \frac{v_D . c_{D,F} . M_S}{c_{S,F} . M_D} \right)}{J_V + \alpha c_{S,F}^{\beta} c^{*-\beta}} \tag{36}$$

For the permeate flux these equations can be used:

$$J_V = A(\Delta P - \Delta \pi_s - \delta) \tag{37}$$

Eq. (37) is the osmotic pressure model. This model is used in similar form by many authors [5,10,12-14,17,18]. Parameter A (water permeability) can be concentration or viscous depended [12,14]. For our model we assume this parameter as constant. Coefficient δ represents the effect of dye on flux. This means mainly osmotic pressure of dye. If this parameter represents only osmotic pressure of dye, then it is constant too (constant dye concentration).

The osmotic pressure gradient for salt is related to the difference of the concentration Δc by the van´t Hoff law:

$$\Delta \pi_S = \frac{\nu R^* T}{M} \Delta c_S \tag{38}$$

where *c* is concentration,

A^* - membrane area,

A - water permeability,

B - salt permeability,

J - flux,

R - rejection,

R^* - universal gas constant,

M - relative molecular mass,

δ - coefficient for dye solution,

σ - reflection coefficient

ν - valence (for NaCl is ν = 2 and for dye ν = 3).

α - coefficient for salt permeability,

β - coefficient for concentration dependence of salt permeability

subscripts

$_S$ - salt

$_D$ - dye

$_V$ - water

$_F$ - feed

$_P$ - permeate

$_R$ - retentate

$_W$ - membrane interface (wall)

$_0$ - beginning of the concentration mode

1.5. Characterization of membranes

Before diafiltration experiments characterizations of commercial membranes are carried out. For these characterizations pure water and water solutions of salt are used. From experiments with pure water model parameter *A* (water permeability) can be estimated. This parameter is slope of the curve (straight line) J = f (ΔP) (see Eq. (37) and $\Delta\pi = \delta = 0$ because no salt and dye are used). In our model we assume water permeability as constant. However, an increase in concentration can cause significant changes in viscosity and a consequent modification of the water permeability. According to resistance model ($A = 1/(\mu\, R_M)$) the dependence of water permeability on viscosity can be expressed as:

$$A_\mu = \frac{A}{\mu_{REL}} \tag{39}$$

where *A* is the water permeability respect to pure water and μ_{REL} is the relative viscosity of feed solution to pure water [12].

In case of diafiltration fouling or gel layer effects can occur and then parameter A is depended on dye/salt concentration ratio (in resistance model is added next resistance $A=1/\mu_{REL}\,(R_M + R_F)$, where R_M and R_F are membrane and fouling resistance [14].

Similar experiments are made with salt solutions. Four salt concentrations (1, 5, 10 and 35 g/l) are used. From these experiments can be obtained parameter B (salt permeability) and then α and β (plotting B versus c_F). Values obtained from these experiments are not used direct but are used as first approximation values for best fit parameters (see Table 3). From results (salt rejection and flux) the suitable membranes for desalination were chosen, Desal 5DK, NF 70, NF 270 and TR 60. Membranes NF 90 and Esna 1 had higher rejection (see Figure 4). For desalination, than membrane with small rejection of salt are suitable.

1.6. Comparison of membranes

For comparison of membranes, three factors were used.

The first factor is separation factor of diafiltration, *S*:

$$S = \frac{\dfrac{c_D}{c_D^0}}{\dfrac{c_S}{c_S^0}} = \frac{c_D c_S^0}{c_D^0 c_S} \tag{40}$$

where c^0_D, c^0_S are concentrations of dye and salt at the beginning of experiment, c_D, c_S are concentrations of dye and salt in the end of experiment.

The separation factor, S, represents how well the dye will be desalinated. With higher separation factor the dye desalination is better. But it is also clear that with bigger separation factor the loss of the dye will be bigger because real membranes have not 100% rejection of dye.

The dye loss factor, Z, can be defined as the rates of amount of the dye in permeate to amount of the dye at the beginning of experiment:

$$Z = \frac{V\left(c_D^0 - c_D\right)}{V c_D^0} = 1 - \frac{c_D}{c_D^0} \tag{41}$$

The third parameter is time of diafiltration needed to reach certain separation factor, S. The total time of diafiltration with n steps, τ, can be expressed (constant permeate flux in each concentration mode) as:

$$\tau_{total} = \sum_{i=1}^{n} \frac{\Delta V}{Q} = \sum_{i=1}^{n} \frac{V_{F0} - V_F}{AJ} \tag{42}$$

where Q is flow of permeate.

2. Methods

2.1. Membranes

Eight NF membranes were chosen for this study. Properties of membranes used are given in Table 1.

Indication	Type	Producer	MWCO [Da]	Material	Module
Desal 5DK	Desal 5DK	GEW & PT	200	polyamide	spiral-wound
Esna 1	Esna 1	Hydranautics	100-300	polyamide	spiral-wound
NF 270	NF 270	Dow	270	polyamide	spiral-wound
NF 90	NF 90	Dow	90	polyamide	spiral-wound
NF 70	CSM NE 2540-70	Saehan	250	polyamide	spiral-wound
NF 45	NF 45	FILMTEC	100	polyamide	spiral-wound
TR 60	TR 60 - 2540	Toray	400	polyamide	spiral-wound
PES 10	PES 10	Hoechst	500-1000	polyether-sulphone	spiral-wound

Table 1. Properties of the membranes used.

2.2. Materials

Dye was obtained from VÚOS a.s. Pardubice, Czech Republic. The commercial name is Reactive Orange 35, and a molecular weight is 748.2 Da in free acid form (three acidic groups) or 817.2 Da as the sodium salt. Figure 2 shows structural formula of the free acid form.

NaCl and $MgSO_4$ used for all experiments were analytical grade. The demineralised water with the conductivity between 4-15 µS/cm was used in this study.

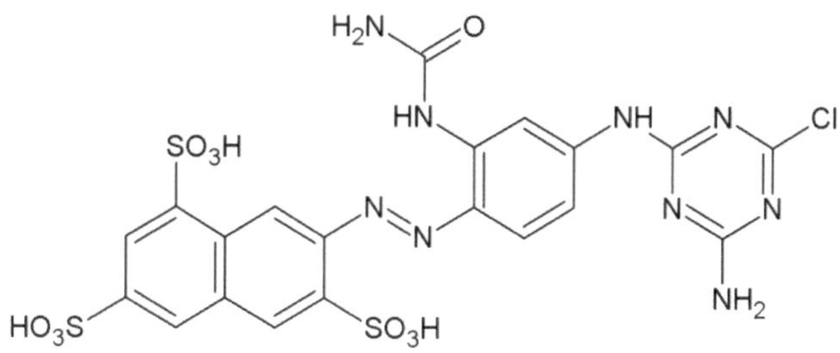

Figure 2. Structural formula of dye (free acid).

2.3. Experimental system

Experiments were carried out on system depicted schematically on Figure 3. Feed (F) was pumped by pump (3) (Wanner Engineering, Inc., type Hydracell G13) from feed vessel (2) to membrane module (1). Pressure was set by valve (4) placed behind membrane module. Permeate (P) and retentate (R) were brought back to feed vessel. Pressure was measured by manometer (5). Temperature was detected by thermometer (6). Stable temperature was maintained by cooling system (7).

2.4. Analytical methods

Dye concentrations were analysed using a spectrophotometer (SPECOL 11). NaCl and $MgSO_4$ concentrations were calculated from conductivity measurements using a conductivity meter (Cond 340i). Permeate and retentate salt concentrations during diafiltration experiments were analysed using potentiometric titration.

2.5. Separation procedure

The system was operated in the full recirculation mode while both retentate and permeate were continuously recirculated to the feed tank except sampling and concentration mode of diafiltration. By changing applied pressure (from 5 to 30 bar) and concentration of salt (1, 5, 10 and 35 g/l) in characterization of membranes both the retentate and permeate were returned back to the feed tank for 0.5 h or 10 min, respectively to reach a steady state before sampling. Before first concentration mode in diafiltraton experiments and after each diluting mode the total recirculation was used 1h and minimally 5 min, respectively. The permeate flux was measured by weighing of certain permeate volume and using a stopwatch.

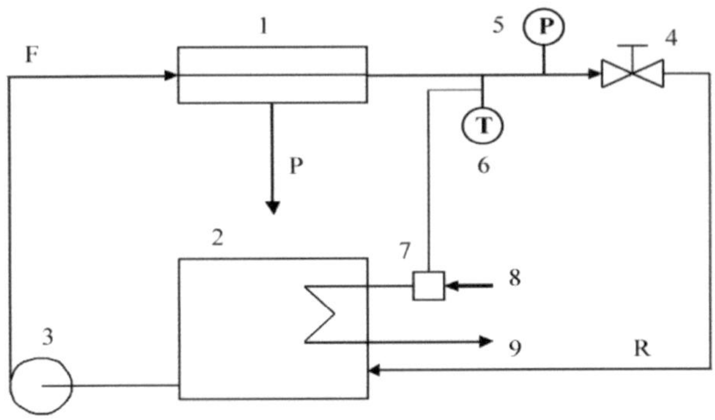

Figure 3. Schematic diagram of the experimental set-up used: 1 membrane module, 2 feed vessel, 3 high pressure pump, 4 back pressure valve, 5 manometer, 6 temperature controller, 7 cooling system, 8 cooling water input, 9 cooling water output, F feed, P permeate, R retentate (concentrate).

3. Results and discussions

3.1. Pure water flux

Water permeability is one of the basic characteristic of NF membranes. The pure water permeability of the eight membranes was determined by measuring the deionized water flux at different operating pressures. According to Darcy´s law, the permeate flux is directly proportional to the pressure difference across the membrane. The slope of this line corresponds to the water permeability (A).

Membrane	A [l/m^2.h.bar]
Desal 5DK	3.365
Esna 1	4.824
NF 90	5.845
NF 270	6.801
NF 70	2.650
NF 45	3.184
TR 60	3.952
PES 10	12.583

Table 2. Water permeability of membranes used.

As we can see the water permeability of PES 10 is approximately three times higher than water permeability of other membranes. It can be due to more open structure of this membrane, which can approach to the UF type. But the opened structure may cause the insufficient retention of the dye in the case of the dye-salt separation.

3.2. Flux and salt rejection in single salt solutions

Basic membrane characteristics are the dependence of the permeate flux and the salt rejection on other operation parameters, i.e. the applied pressure difference and the salt concentration in feed.

The permeate flux increases with increasing pressure and decreases as the feed concentration of salt increases. For the lowest concentration of salt (1 g/l), the values of permeate flux were similar to the values of clean water. The lower values of permeate flux were obtained with the increasing salt concentrations in feed (increasing osmotic pressure). For membrane NF 90 fluxes were not measured at the highest salt concentration for pressure smaller than 25 bar, because the osmotic pressure was too high. Opposite problem was with membrane NF 270 at the smallest salt concentration in feed. The permeate flux was too high and pump was not able to deliver necessary volumetric flow of retentate (600 l/h) for constant conditions at all experiments.

The observed rejection increases as the pressure difference increases, and decreases with the increasing salt concentration in feed for all tested membranes. However, the minimal values were obtained during experiments with membrane NF 270. Low values of the salt rejection and higher values of the permeate flux are suitable for desalting. Figure 4 shows the comparison of tested membranes for the lowest (1 g/l) and the highest salt concentrations in feed (35 g/l), respectively.

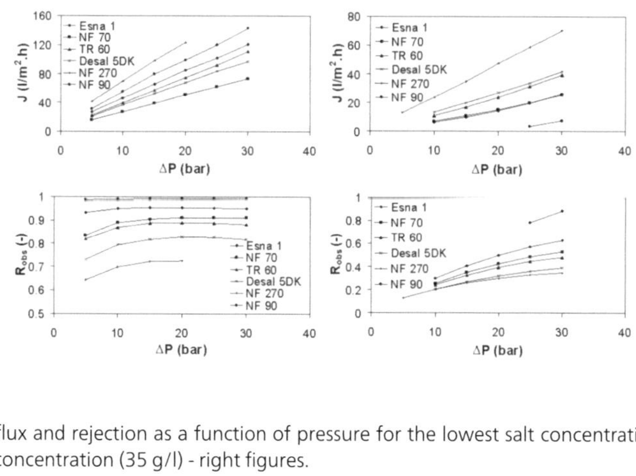

Figure 4. Permeate flux and rejection as a function of pressure for the lowest salt concentration; (1 g/l) - left figures and the highest salt concentration (35 g/l) - right figures.

Figure 5 denote dependence of the transmembrane pressure on the flux for NaCl solutions with the various salt contents. As we can see this dependence for every salt concentration shows a straight line course. Thus we can assume that the concentration polarization has no significant influence and therefore we can consider the bulk concentration equal to that on the membrane [20], which is required in model equations.

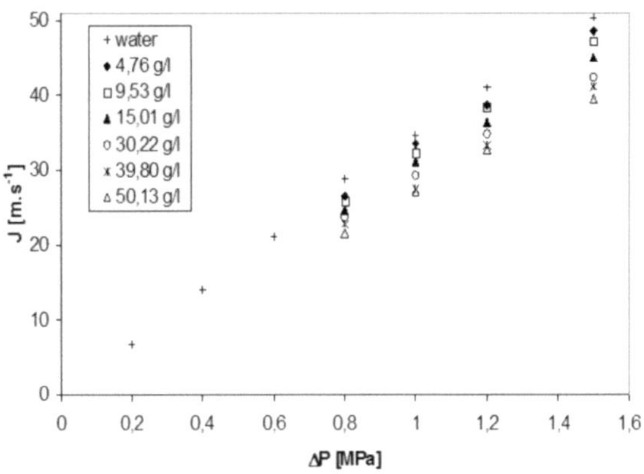

Figure 5. Flux as a function of transmembrane pressure for NaCl solutions with various salt content (Desal 5DK).

Experimental data rejection-flux can be evaluated by extended Spiegler-Kedem model to obtain parameters σ, α and β.

Membrane	L_p [m.s^{-1}.Pa^{-1}]	σ [-]	α [m.s^{-1}]	β [-]
Desal 5DK	$9.35.10^{-12}$	0.824	$1.96.10^{-6}$	0.438
Esna 1	$1.34.10^{-11}$	0.594	$1.60.10^{-6}$	0.620
NF 90	$1.62.10^{-11}$	0.685	$4.39.10^{-6}$	0.266
NF 270	$1.89.10^{-11}$	0.723	$2.88.10^{-6}$	0.529
NF 70	$7.36.10^{-12}$	0.598	$1.34.10^{-6}$	0.381
TR 60	$1.10.10^{-11}$	0.583	$1.06.10^{-6}$	0.471

Table 3. Coefficients of extended Spiegler-Kedem model for the transport of NaCl for various membranes.

The membrane PES 10 was not involved into Table 3 because this membrane didn't show typical course of the rejection-flux dependence. These differences are caused by another nature of this membrane near the UF type and therefore the transport through this membrane cannot be described by Spiegler-Kedem model.

On the Table 3 we can see that the water permeability and the salt permeability show the same trend for various membranes (L_P and α). The parameter σ means the maximum rejection attainable on given membrane (at the lowest concentration and the highest pressure). The parameter β express the concentration dependence of the salt permeability. The analyzation of experimental data for separation of $MgSO_4$ by Spiegler-Kedem model is less interesting because except PES 10 all membranes showed rejection almost equal unity.

3.3. Flux, salt and dye rejection in mixed dye - salt solutions

The aim of these experiments was to find dependence of salt and dye rejections on the salt and dye concentration. In every experiment carried out in this work the dye rejection was sufficient high and almost equal unity. The lowest value of the dye rejection observed was 0.9988. The salt rejection as a function of the dye and salt concentration is plotted in following Figure 6.

It can be seen from Figure 6 that the salt rejection decreases with decreasing salt concentration and with increasing dye concentration, corresponding to Donnan equilibrium (Eq. 15).

In the case of solution without the dye or with low dye content (positive rejections) we can observe the typical decline of salt rejection with increasing salt concentration. At higher dye content we can observe increase of salt rejection with increasing salt concentration due to shifting of Donnan equilibrium.

It's obvious from Figure 7 that salt content has also a strong influence on the flux. In the case of single salt solution (without dye content) we can see a typical decrease of flux with increasing salt content. In the case of mixed dye-salt solutions we can observe initial increase of flux and following decrease after a maximum was reached. The initial increase of flux at high dye concentration and low salt concentration is due to negative rejection (see Figure 6), which causes the reverse osmotic pressure difference between permeate and feed side of the membrane ($\Delta\pi>0$). This reverse osmosis pressure difference escalates the driving force of the process (Eq. 1) thus flux increases.

The experimental dependence of the salt rejection on flux can be evaluated by extended Spiegler-Kedem model (Figure 8) in order to obtained parameters σ_{NaCl}, α_{NaCl} and β_{NaCl}. These parameters are characteristic for the transport of NaCl through given membrane and also characteristic for given dye in feed solution dye. The meaning of individual parameters is the same as in the case of the single salt transport.

Figure 8 depicted the experimental dependence of the salt rejection on the flux. The single curves represent course of the salt rejection as a function of the flux for given dye content in the feed. The pressure difference was kept constant during all experiments and flux was changed by changing of salt content in the feed (changing of osmotic pressure difference).

It can be seen from Figure 8 that extended Spiegler-Kedem model isn't able to evaluate rejection-flux data as accurately as it was in the case of the single salt transport. But realising the range of rejection values we can consider prediction by this model still sufficient.

We can see that the coefficient α is approximately four times higher than in the case of the separation of single NaCl solution, which reflect the fact that the presence of the dye escalate the salt permeability through the membrane.

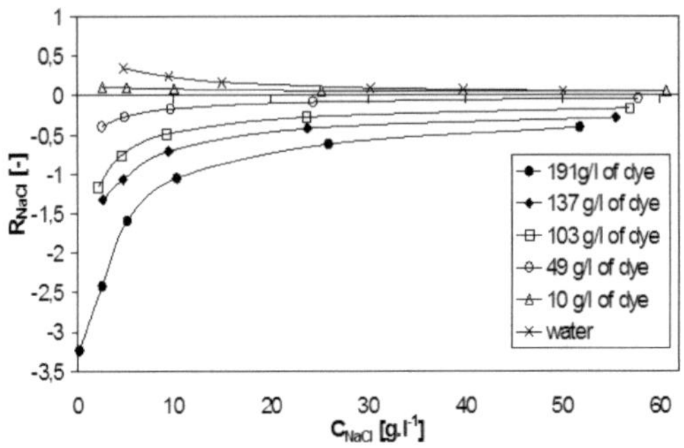

Figure 6. Salt rejection as a function of salt concentration for different dye concentration (Desal 5DK; 1.5MPa).

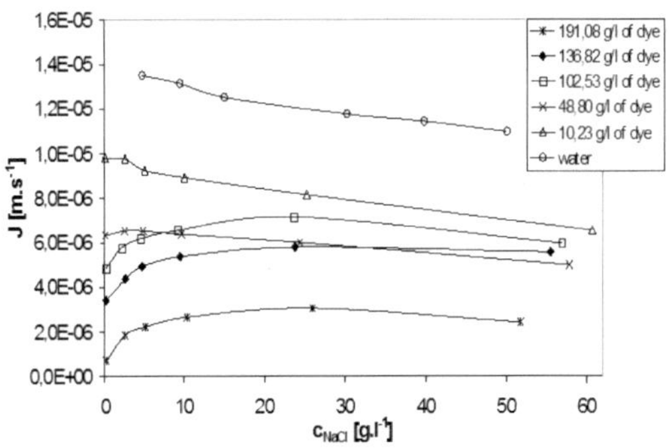

Figure 7. Flux as a function of salt concentration for different dye concentration (Desal 5DK).

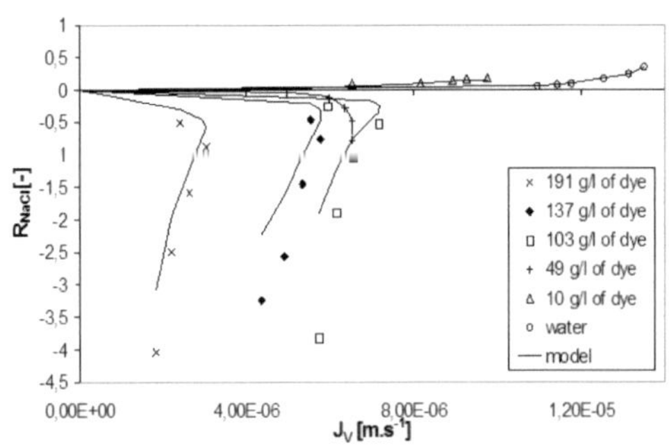

Figure 8. Salt rejection as a function of flux for different dye concentration (Desal 5DK; 1.5 MPa); σ_{NaCl}=0.880; α_{NaCl}=5.38.10^{-6} and β_{NaCl}=0.623.

http://dx.doi.org/10.5772/50438

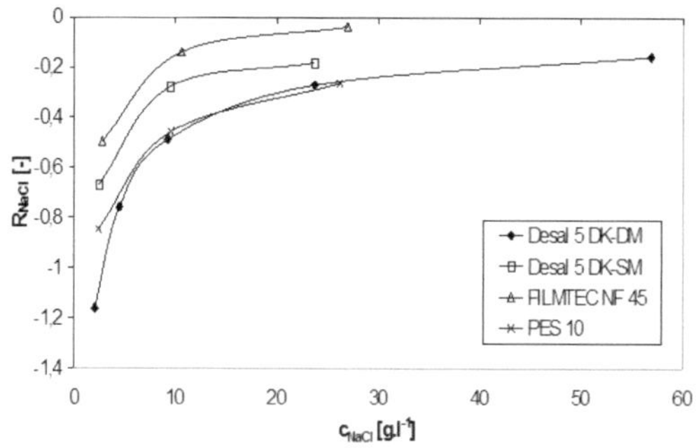

Figure 9. Comparison of rejection-NaCl concentration dependence for various membranes at dye concentration 100 g.l^{-1} (1.5 MPa).

The comparison with another membranes was carried out in order to determine the most suitable one of given membranes. As we can see at Figure 9 membranes Desal 5DK and PES 10 have the best course of dependence of salt rejection as a function of concentration but PES 10 shows the highest flux (Figure 10). Among given membranes PES 10 is therefore the most suitable membrane for desalination of this dye.

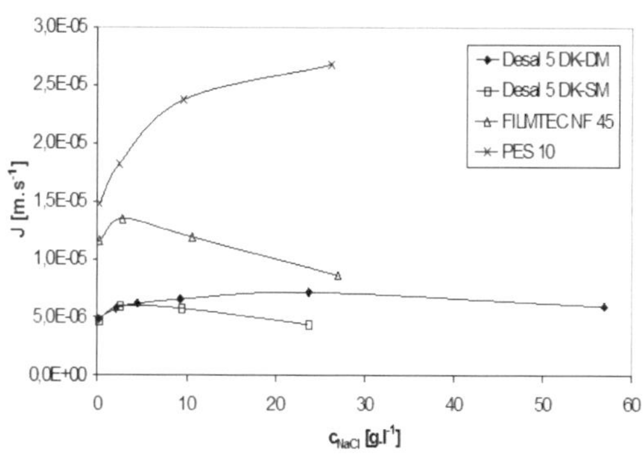

Figure 10. Comparison of flux-NaCl concentration dependence for various membranes at dye concentration 100 g.l^{-1} (1.5 MPa).

3.4. Diafiltration

The concentration of macrosolutes by batch NF is frequently accompanied by a diafiltration step to remove microsolutes such as salts. Batch diafiltration with periodically adding solvent at 20 bars and constant retentate flow 600 l/h was provided. Aqueous dye solutions with dye concentrations 100, 50 and 10 g/l and salt concentration between 20-23 g/l were desalted at 23 C. Volume of the pure solvent added in every dilute mode was 4l (the same volume of permeate was remove before in concentration mode). Total feed volume in tank was 52l. For every membrane and every concentration of the dye in feed fifty diafiltration steps were made. One point in Figures 11-14 is one diafiltration step before concentration mode.

Four membranes only - Desal 5DK, NF 70, NF 270 and TR 60 - were used for diafiltration experiments. Diafiltration with membrane NF 270 was provided only with dye concentrations 100 and 50 g/l and with membrane TR 60 only at the highest dye concentration, which is the best for desalination. For the reason of low values of permeate flux, membrane NF 90 and Esna 1 were not used for those experiments.

Dependences of rejection on salt concentration in feed are given in Figure 11 for Desal 5DK, NF 70, NF 270 and TR 60, respectively. Membranes are compared at dye concentration 100 g/l. The lowest values of rejection (max. 0.29) were obtained for membrane Desal 5DK. The membrane NF 70 had the highest values.

Dependences of flux on salt concentrations are shown in Figure 12. The highest values of flux (70.3 $l/m^2.h$) were obtained in experiments with membrane NF 270. The permeate flux decreased while salt concentration increased. This is due to the effect of osmotic pressure along with the concentration polarization. Due to the concentration polarization phenomenon, the osmotic pressure of the aqueous solution adjacent to the membrane active layer is higher than the corresponding value of the feed solution. As a result, the osmotic pressure would increase dramatically while the salt concentration increased.

In Figure 13 dependences of salt concentrations on time of diafiltration are shown.

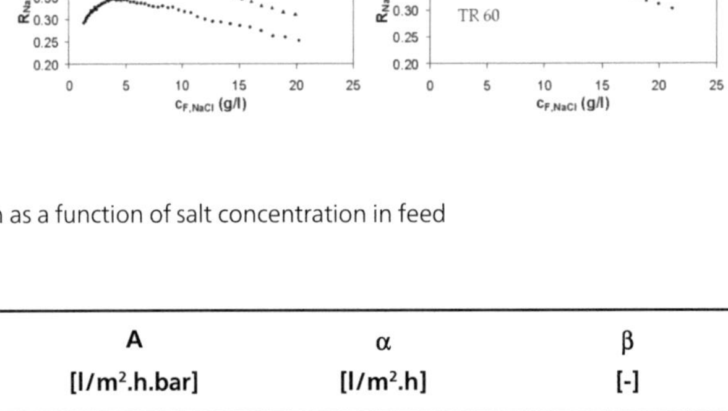

Figure 11. Salt rejection as a function of salt concentration in feed

Membrane	A [$l/m^2.h.bar$]	α [$l/m^2.h$]	β [-]	δ [bar]
Desal 5DK	3.365	5.379	0.623	7.503
NF 70	2.650	4.839	0.381	4.664
NF 270	6.801	10.350	0.529	8.561
TR 60	3.952	6.693	0.474	5.888

Table 4. Model parameters.

The comparisons of experimental and model data for the highest dye concentration (100 g/l) are shown in Figure 14. Salt concentrations are calculated using Eq. (28) and Eq. (30). Rejection needed for these equations is calculated on basis of Eq. (36). Best fit parameters for proposed model are given in Table 4.

From Table 4 can be shown that δ is not only osmotic pressure (if we assume water permeability as constant), because the values of δ are different. From these results we can assume, the highest effect of dye on flux is for membrane NF 270. This membrane is the most fouled from these membranes. It is appropriate assumed the change in water permeability (A) in case of desalination of dyes.

In our experiments, the decrease of permeate flux was mainly caused by the effect of concentration polarization and the increase of the viscosity of dye solution. The dye formed a boundary layer over the membrane surface (concentration polarization) and consequently, increased the resistance against the water flux through the membrane. At the same time, the viscosity of solution increased with higher concentration.

From Figure 14 can be shown that the experimental results in permeate fit the model very well. Due to considerably low salt concentrations in permeate, concentration polarization was minimized. The diafiltration process benefits to obtain pure salt product and this can be predicted by a mathematic model on the basis of description of discontinuous diafiltration by periodically adding solvent at constant pressure difference.

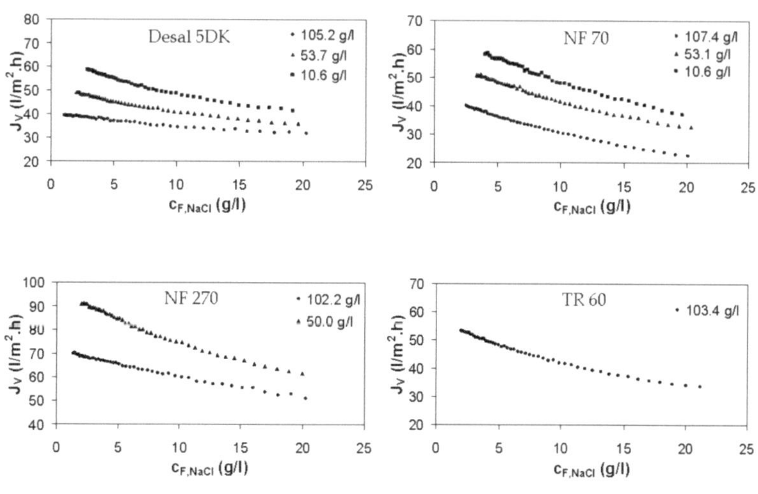

Figure 12. Permeate flux as a function of salt concentration in feed.

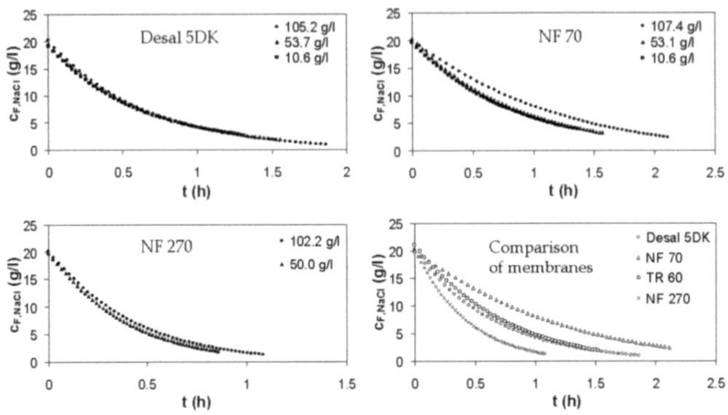

Figure 13. Salt concentration as a function of diafiltration time for membranes Desal 5DK, NF 70, NF 270 and comparison of tested membranes at dye concentration 100 g/l.

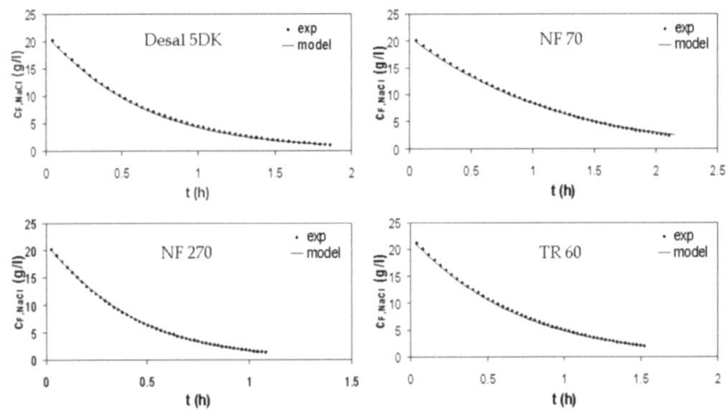

Figure 14. Comparison of experimental and model data for the highest dye concentration.

From Table 5 is clearly shown, the total time of diafiltration, τ_{total}, decreases with decreasing dye concentration. The shortest time had membrane NF 270. Time for the highest dye concentration is not two times higher than with medium dye concentration for all tested membranes (the time/amount of dye desalted ratio is the smaller for higher concentration of dye). Separation factor decreases with decreasing concentration of dye and it is the second reason why the highest dye concentration was used as the best mode for desalination. The best separation factor had membrane Desal 5DK (very similar values, except the highest dye concentration, had membrane NF 270). The loss of dye is almost same for membrane Desal 5DK, NF 70 and NF 270 at all concentrations of dye. Only for membrane TR 60 are obtained higher loss of dye.

		DESAL 5DK		
$c_{F,NaCl,Z}$	(g/l)	22.00	19.62	19.21
$c_{F,NaCl,K}$	(g/l)	0.97	1.88	2.70
$c_{F,dye,Z}$	(g/l)	105.17	53.75	10.61
$c_{F, dye,K}$	(g/l)	105.15	53.74	10.60
τ_{total}	(hod)	1.86	1.55	1.32
S	(-)	22.71	10.44	7.10
Z	(%)	0.02	0.01	0.07
		NF 70		
$c_{F,NaCl,Z}$	(g/l)	20.08	20.33	19.64
$c_{F,NaCl,K}$	(g/l)	2.40	3.21	3.86
$c_{F, dye,Z}$	(g/l)	107.42	53.05	10.56
$c_{F, dye,K}$	(g/l)	107.38	53.04	10.56
τ_{total}	(hod)	2.11	1.57	1.37
S	(-)	8.37	6.34	5.09
Z	(%)	0.03	0.03	0.05
		TR 60		
$c_{F,NaCl,Z}$	(g/l)	21.14	20.56	19.72
$c_{F,NaCl,K}$	(g/l)	1.88	2.22	3.16
$c_{F, dye,Z}$	(g/l)	103.41	53.65	10.71
$c_{F, dye,K}$	(g/l)	102.74	53.16	10.52
τ_{total}	(hod)	1.52	1.25	0.98
S	(-)	11.17	8.98	6.25
Z	(%)	0.65	0.48	0.18
		NF 270		
$c_{F,NaCl,Z}$	(g/l)	22.00	19.92	20.64
$c_{F,NaCl,K}$	(g/l)	1.28	1.90	2.68
$c_{F, dye,Z}$	(g/l)	102.15	49.98	10.52
$c_{F, dye,K}$	(g/l)	102.12	49.97	10.53
τ_{total}	(hod)	1.08	0.86	0.67
S	(-)	17.19	10.50	6.89
Z	(%)	0.03	0.04	0.04

Table 5. Total time of diafiltration, τ_{total}, the separation factor, **S**, and the loss of dye, **Z**. (subscript $_{Z,K}$ are start and end of diafiltration, respectively)

4. Conclusions

The separation performance of dye, salt and dye solution with six different nanofiltration membranes were investigated, followed by the study of the optimum of diafiltration and concentration process of dye solution.

Asymmetric and negatively charged polyamide thin-film composite membranes of near similar molecular weight cut-off were characterized for key physical and surface properties, and employed to perform the laboratory-scale experiments to investigate the impacts of membranes properties on reactive dye removal from dye/salt mixtures through NF process. It was found that properties of the NF membrane play an important role in dye removal rate, stable permeate flux and their change behaviour with operational conditions.

The electrostatic repulsive interaction between dye and membrane surface promotes the dye removal and decreases concentration polarization and dye adsorption on the membrane surface. But, the action will be weakened as the dye concentration or salt concentration increased.

The introduction of an exponential term for the concentration dependence of salt permeability in the Spiegler-Kedem model allows very good prediction of rejection of nanofiltration membranes for single salt solutions depending on the feed concentration and permeates flux.

In the case of separation of mixed dye-salt solutions the extended Spiegler-Kedem model including Donnan equilibrium term (the Perry-Linder model) and the exponential concentration dependence term can be used for sufficient prediction of the salt rejection even at high dye concentrations typical for industrial desalination process.

From the results presented above it is clear that the best concentration of the dye in feed for desalination of reactive dye by batch diafiltration is 100 g/l. In this case the salt rejection reaches minimal value due to Donnan potential which strengthens the flow of salt through the membrane.

The best membrane for desalination is NF 270 which has smaller dye loss factor and the shortest time of diafiltration. Very suitable membrane is also Desal 5DK, which has the best separation factor and dye loss factor, but this membrane has longer time of diafiltration (see Table 5). For desalination qualitative description it is convenient to use the proposed model.

Acknowledgements

This project was financially supported by Ministry of Education, Youth and Sports of the Czech Republic, Project SGFChT05/2012.

http://dx.doi.org/10.5772/50438

Author details

Petr Mikulášek* and Jiří Cuhorka

*Address all correspondence to: petr.mikulasek@upce.cz

Institute of Environmental and Chemical Engineering, University of Pardubice, Pardubice, Czech Republic

References

[1] Drioli, E., Laganh, F., Criscuoli, A., & Barbieri, G. (1999). Integrated Membrane Operations in Desalination Processes. *Desalination*, 122(2-3), 141-145.

[2] Diawara, C. K. (2008). Nanofiltration Process Efficiency in Water Desalination. *Separation & Purification Reviews*, 37(3), 303-325.

[3] Yu, S., Gao, C., Su, H., & Liu, M. (2001). Nanofiltration used for Desalination and Concentration in Dye Production. *Desalination*, 140(1), 97-100.

[4] He, Y., Li, G. M., Zhao, J. F., & Su, H. X. (2007). Membrane Technology: Reactive Dyes and Cleaner Production. *Filtration & Separation*, 44(4), 22-24.

[5] Mulder, M. (2000). Basic Principles of Membrane Technology. 2nd Ed., Dordrecht, Kluwer Academic Publishers.

[6] Oatley, D. L., Cassey, B., Jones, P., & Bowen, W. R. (2005). Modelling the Performance of Membrane Nanofiltration - Recovery of high-value Product from a Process Waste Stream. *Chemical Engineering Science*, 60(7), 1953-1964.

[7] Bowen, W. R., & Mohammad, A. W. (1998). Diafiltration by Nanofiltration: Prediction and Optimization. *AIChE Journal*, 44(8), 1799-1812.

[8] Weselowska, K., Koter, S., & Bodzek, M. (2004). Modelling of Nanofiltration in Softening Water. *Desalination*, 162(1-3), 137-151.

[9] Foley, G. (2006). Water Usage in Variable Volume Diafiltration: Comparison with Ultrafiltration and Constant Volume Diafiltration. *Desalination*, 196(1-3), 160-163.

[10] Al-Zoubi, H., Hilal, N., Darwish, N. A., & Mohammed, A. W. (2007). Rejection and Modelling of Sulphate and Potassium Salts by Nanofiltration Membranes: Neural Network and Spiegler-Kedem Model. *Desalination*, 206(1-3), 42-60.

[11] Wang, L., Yang, G., Xing, W., & Xu, N. (2008). Mathematic Model of Yield for Diafiltration. *Separation and Purification Technology*, 59(2), 206-213.

[12] Kovács, Z., Discacciati, M., & Samhaber, W. (2009). Modelling of Batch and Semibatch Membrane Filtration Processes. *Journal of Membrane Science*, 327(1-2), 164-173.

[13] Das, Ch., Dasgupta, S., & De , S. (2008). Steady-state Modelling for Membrane Separation of Pretreated Soaking Effluent under Cross Flow Mode. *Environmental Progress*, 27(3), 346-352.

[14] Cséfalvay, E., Pauer, V., & Mizsey, P. (2009). Recovery of Copper from Process Waters by Nanofiltration and Reverse Osmosis. *Desalination*, 240(1-3), 132-146.

[15] Hussain, A. A., Nataraj, S. K., Abashar, M. E. E., Al-Mutaz, I. S., & Aminabhavi, T. M. (2008). Prediction of Physical Properties of Nanofiltration Membranes using Experiment and Theoretical Models. *Journal of Membrane Science*, 310(1-2), 321-336.

[16] Schirg, P., & Widmer, F. (1992). Characterisation of Nanofiltration Membranes for the Separation of Aqueous Dye-salt Solution. *Desalination*, 89(1), 89-107.

[17] Koyuncu, I., & Topacik, D. (2002). Effect of Organic Ion on the Separation of Salts by Nanofiltration Membranes. *Journal of Membrane Science*, 195(2), 247-263.

[18] Kovács, Z., Discacciati, M., & Samhaber, W. (2009). Modelling of Amino Acid Nanofiltration by Irreversible Thermodynamics. *Journal of Membrane Science*, 332(1-2), 38-49.

[19] Spiegler, K. S., & Kedem, O. (1966). Thermodynamics of Hyperfiltration (Reverse Osmosis): Criteria for Efficient Membranes. *Desalination*, 1(4), 311-326.

[20] Xu, Y., & Lebrun, R. E. (1999). Comparison of Nanofiltration Properties of Two Membranes using Electrolyte and Non-electrolyte Solutes. *Desalination*, 122(1), 95-105.

Chapter 8

Desorption of Cadmium from Porous Chitosan Beads

Tzu-Yang Hsien and Yu-Ling Liu

Additional information is available at the end of the chapter

http://dx.doi.org/10.5772/50142

1. Introduction

Cadmium pollution of the environment has become a serious problem due to the increasing consumption of cadmium by industry in the past 20 years. Cadmium is introduced into the environment from the effluence of electroplating industry, and in solid and aqueous discharges from mining operations. Increased environmental awareness has resulted in the promulgation of more stringent legislation in several countries for water quality. For example, in Italy and the United States, the maximum permitted concentrations of heavy metals such as cadmium, lead, chromium, and nickel ions are 5, 50, 50, and 50 μg/l respectively [1,2]. In order to increase the concentration of exposed active sites within chitosan so that the adsorption capacity and transport rate of metal ions into the particle can be enhanced, porous chitosan beads need to be developed. Rorrer, Way, and Hsien [3]described the synthesis of 1 mm and 3 mm porous, magnetic, chitosan beads for cadmium ion separation from aqueous solutions. Complete adsorption isotherms over a large range of cadmium ion concentrations (2-1700 ppm) onto the chitosan beads were obtained. Chitosan is a cationic polymer which can displace adsorbed metal ions by hydrogen ions in a low pH environment. Muzzarelli et al. [4]pointed that a packed column of mercury-adsorbed chitosan could be regenerated by flushing the bed with a 10 mM potassium iodide solution or other eluting agents. Randall et al. [5] regenerated chitosan powder in a packed column by flushing the bed with a 0.2 N NH4Cl solution. Nickel removal efficiencies were as high as 97 %. The pH effect on the desorption process for regeneration of cadmium-adsorbed chitosan powder was first considered by Jha et al. [6]. An economic comparison of two different processes to recycle chitosan after the adsorption process was provided by Coughlin et al. [7]. Thus, it is valuable to define the optimum regeneration parameters for chitosan beads. The recycle efficiency of the cadmium-adsorbed chitosan beads will also be considered by this study.

The hydrogen ion consumption and optimum pH range for the regeneration of chitosan beads after cadmium adsorption will be studied. Specifically, the desorption process after cadmium adsorption will be carried out in a spinning-basket reactor with online pH measurement to determine the effect of pH and H+ consumption on cadmium desorption and bead regeneration efficiency. The objective of this study is to determine the feasible pH range and H+ consumption for regeneration of cadmium adsorbed chitosan beads. The pH affects cadmium adsorption and desorption on the chitosan beads. In this present study, the adsorption and desorption kinetics for single stage adsorption desorption experiments are determined in a spinning basket reactor. The hydrogen ion capacity of the chitosan beads and the pH of the cadmium solution in the vessel are also measured as a function of time. Finally, an equilibrium model for desorption process is presented in order to describe the competitive relationships associated with displacing adsorbed cadmium ions with hydrogen ions.

2. Materials and Methods

2.1. Chitosan Beads Synthesis

The synthesis of chitosan beads, including chitosan solution preparation, gel beads casting, crosslinking and freeze drying was described in previous studies [3,8]. Specifically, a 5 wt % chitosan solution was casted in the precipitation bath to form chitosan gel beads and then crosslinked with a 2.5 wt % aqueous glutaric dialdehyde. The wet crosslinked chitosan beads were freeze-dried to remove the remaining humidity. The chitosan beads crosslinked with a 2.5 wt % initial glutaric dialdehyde solution were used in this study.

2.2. Single Stage Desorption

Two types of desorption experiments, single stage and multiple stage, were performed. In the single stage experiment, the adsorbed cadmium ions on chitosan beads were released back to the bulk solution by the single addition of a large amount of H^+. Batch adsorption/desorption experiments were conducted at 25 ºC in a spinning-basket reactor (Figure 1), inspired by a basket reactor from Carberry [9]. Prior to desorption, the adsorption process was carried out. Specifically, 0.5 g of chitosan beads were packed into the hollow impeller basket assembly and contacted with 200 mL of 200 mg /L cadmium ion solution at 150 rpm and 25ºC for at least 48 hours to ensure that adsorption equilibrium was achieved. After adsorption, 65 mL of 0.1 N HNO_3 solution was added to the vessel to load the bulk solution with the H^+ ions needed to affect the complete desorption and to reach a final pH value of 2.0. The adsorption/desorption parameters for spinning basket reactor experiments are summarized in Table 1.

The high concentration of hydrogen ions loaded to the vessel displaced the adsorbed cadmium ions. The kinetics of cadmium release and H^+ adsorption were followed by measuring the cadmium ion concentration and pH of the bulk solution phase with time. The pH was monitored continuously, whereas 0.5 mL samples were periodically removed from the reactor and analyzed for Cd^{+2} concentration by ion chromatography (IC).

http://dx.doi.org/10.5772/50142

Process Condition	Variable and Units
Temperature	25 ºC
Bead loading (m_b)	0.5 g
Initial solution volume (V)	200 mL
Initial Cd^{+2} concentration (C_o)	200 mg Cd/L
Acid (HNO_3) concentration (C_a)	0.0001 - 0.1 N
Agitation	150 rpm

Table 1. Table 1. The adsorption/desorption parameters for spinning basket reactor experiments

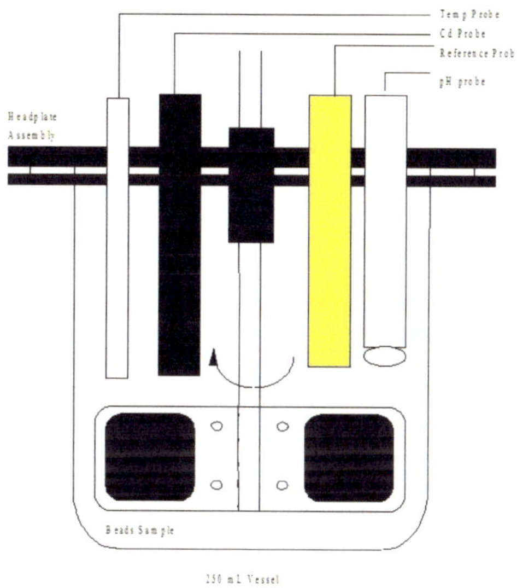

Figure 1. Spinning basket reactor

Two control experiments were performed. In the first control experiment, the pH curve for the chitosan beads was measured to determine the extent of H^+ consumed by adsorption for the case where no cadmium ions were present in the system. Specifically, 0.5 g of chitosan beads were packed in the hollow propeller cells of the spinning-basket reactor and immersed in 200 mL of deionized distilled water for 48 hours. The pH at the start of the experiment was about 7.0. The mixing speed was 150 rpm. Then, 65 mL of 0.1 N HNO_3 solution was added to the vessel and the pH was measured continuously until an equilibrium pH value was reached. During this time, the pH of the bulk solution increased as hydrogen ions adsorbed onto weakly basic $-NH_2$ sites on chitosan.

In the second control experiment, the stability of chelated cadmium ions adsorbed on chitosan beads was tested. Specifically, the cadmium adsorbed chitosan beads were immersed in 200 mL distilled water within the reactor at 25 º C and mixed at 150 rpm. Samples were removed from bulk solution at 123 and 171 hours and analyzed for cadmium ion concentration by ion chromatography (IC).

Shake flask experiments were carried out to determine the equilibrium hydrogen and cadmium loading on the chitosan beads at different pH levels. Specifically, 0.1 g of chitosan beads were mixed with 40 mL of 200 mg /L cadmium ion solution in a 125 mL Erlenmeyer flask and agitated at 120 rpm and 25 °C for 51 hours to ensure that adsorption equilibrium was reached. After adsorption, different doses (1 to 4 mL) of HNO_3 solution (10^{-4} to 1 N) was added to the flask to process the desorption for another 140 hours. The adsorption/desorption parameters for shake flask experiments are summarized in Table 2.

Process Condition	Variable and Units
Temperature	25 °C
Bead loading (m_b)	0.1 g
Initial solution volume (V)	40 mL
Initial Cd^{+2} concentration (C_o)	200 mg Cd/L
Acid (HNO_3) concentration (C_a)	0.0001 - 0.1 N
Agitation	120 rpm

Table 2. Table 2. The adsorption/desorption parameters for shake flask experiments

2.3. Multiple Stage Desorption

In the multiple stage experiment, adsorbed cadmium was gradually displaced by the series addition of small doses of H^+. A discrete volume of 0.1 N HNO_3 was added to the vessel. If the equilibrium pH was above 2.0, then additional 0.1 N HNO_3 was added. The experiment was completed when an equilibrium pH of 2.0 was achieved.

A modified titration procedure suggested by previous works for studying ion exchange or adsorption characteristics was modified in the present study to determine hydrogen ion adsorption on chitosan beads. Specifically, a 0.5 g of chitosan beads were packed into the hollow propeller cells of the spinning-basket reactor and immersed in 200 mL of deionized distilled water at 150 rpm for 48 hours. After the chitosan beads were saturated with water, a discrete volume of 0.1 N HNO_3 was added to the vessel to initiate hydrogen ion adsorption on chitosan beads. Until an equilibrium pH value was reached, a different amount of 0.1 N HNO_3 solution was sequentially added to the vessel.

2.4. pH and Cadmium Ion Concentration Measurements

The pH measurements were performed with an Orion model 91-02 combination pH electrode and Orion model 720A ion selective electrode (ISE) meter. The pH readings were recorded by a computer every ten minutes during the experiment. The pH saturation curve was expressed as the amount of hydrogen ion sorbed per gram of chitosan beads versus the equilibrium pH of the bulk solution. The cadmium ion concentration was measured by IC analysis as described earlier.

The cadmium adsorption capacity is calculated by

$$Q = \frac{(C_0 - C(t))V}{m_b} \tag{1}$$

where C(t) is the cadmium ion concentration in the vessel at different adsorption or desorption times (mg Cd/L), C_0 is the initial cadmium concentration in the vessel before adsorption (mg Cd/L), m_b is the mass (g) of chitosan beads in the spinning basket reactor impeller assembly, Q is the cadmium adsorption capacity on the chitosan beads (mg Cd/g-chitosan), and V is the current cadmium solution volume loaded in the spinning basket reactor vessel (L).

The percentage of cadmium desorbed from the beads is calculated by

$$D(\%) = \left(1 - \frac{C(t) - C_{min}}{C_0 - C_{min}}\right) \bullet 100 = \left(1 - \frac{Q(t)}{Q_f}\right) \bullet 100\ \% \tag{2}$$

or

$$D(\%) = \left(1 - \frac{C(t) - C_{min,i}}{C_{0,i} - C_{min,i}}\right) \bullet 100 = \left(1 - \frac{Q_i(t)}{Q_{i,f}}\right) \bullet 100\ \% \tag{3}$$

where D (%) is the percentage of cadmium desorbed at a given desorption time, C_{min} is the lowest cadmium concentration along the adsorption or desorption process (mg Cd/L), $C_{0,i}$ is the initial cadmium concentration in the spinning basket reactor at specific desorption stage " *i* " (mg Cd/L), and $C_{min,i}$ is the lowest cadmium concentration along the adsorption or desorption process at a specific desorption stage (mg Cd/L), Q_f is the final cadmium adsorption capacity on the chitosan beads (mg Cd^{+2}/g chitosan).

The hydrogen ion adsorption capacity ($\Delta Q_i(H^+)$) on the chitosan beads was based on pH measurements of the bulk solution in the vessel. The hydrogen ion adsorption capacity at ith desorption stage $\Delta Qi(H^+)$ is given by

$$\Delta Q_i(H^+) = \frac{C_a V_a - (10^{-pH_t} \bullet V_t)}{m_b} \tag{4}$$

where C_a is the concentration of HNO_3 (mole H^+/L) added into the spinning basket reactor before desorption, pH_t is the pH value at different adsorption or desorption times, $\Delta Q_i(H^+)$ is the hydrogen ion adsorption capacity for chitosan beads at the *i*th desorption stage (mg H/g-chitosan), V_a is the acid volume (L) added into the spinning basket reactor, V_t is the volume (L) of solution in the spinning basket reactor at a given time.

The accumulated hydrogen ion adsorption capacity $Q(H^+)$ is given by

$$Q\,(H^+) = \sum_{i=1}^{n} \Delta\, Q_i \tag{5}$$

where n is the number of the desorption stage, $Q(H^+)$ is the accumulated hydrogen ion adsorption capacity in the present stage (mg Cd/g-chitosan).

3. Results and Discussion

3.1. Kinetics of Adsorption and Desorption

The cadmium ion concentration versus time profile in the spinning basket reactor is presented in Figure 2 (a) for a single stage adsorption/desorption experiment. The cadmium ion concentration decreased significantly from 200 to 183 mg Cd^{+2}/L during the first four hours of adsorption and then slightly decreased to reach a final cadmium concentration of 179 mg Cd^{+2}/L at 51 hours. The total solution volume was 196 mL. After adsorption, 65 mL of 0.1 N nitric acid was added to the vessel to initiate the desorption process. The cadmium concentration increased sharply during the first 12 hours of desorption following the addition of nitric acid, and then leveled off after 75 hours of desorption. The hydrogen ion capacity for the chitosan beads and the pH change in the cadmium solution are provided in Figure 2 (b) for both the adsorption and desorption processes. Similarly, the hydrogen ion capacity gradually increased and reached a final value between 8.6 and 9.5 mmole H^+/g chitosan. The pH of the cadmium solution also decreased with the addition of nitric acid and then rose to a final value of 2.17.

Time (h)	Q (mg Cd/g)	pH	Cd^{+2} concentration (mg Cd^{+2}/L)
51	9.86	6.73	0.00
87	9.55	6.51	0.78
144	9.18	6.20	1.71

Table 3. Table 3. Stability of the adsorbed cadmium ion in distilled water at pH 6.0

The stability of the adsorbed cadmium ions in the chitosan beads was determined. Specifically, the cadmium-adsorbed chitosan beads were immersed in 200 mL of distilled water within the spinning basket reactor at 25 ºC and mixed at 150 rpm for 93 hours. Only a trace amount of cadmium ions were released back into the water during the desorption (Table 3). Form this control experiment, we conclude that the exchange

http://dx.doi.org/10.5772/50142

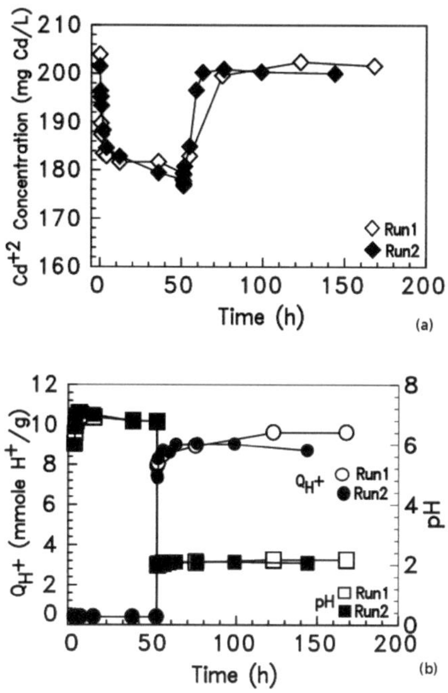

Figure 2. a) cadmium ion concentration versus time (b) hydrogen ion capacity and the pH change in the cadmium solution in the spinning basket reactor for a single stage adsorption/desorption experiment

between the adsorbed cadmium ions with the hydrogen ions on active sites (-NH_2 groups) of the chitosan beads during the desorption process requires hydrogen ions from the addition of the nitric acid.

In another control experiment using the spinning basket reactor, 65 mL of 0.1 N nitric acid were added to 0.5 g chitosan beads to determine the hydrogen ion capacity in cadmium-free water. Profiles for the hydrogen ion consumption and pH are presented in Figure 3. From Figure 3, 8.9 mmole H^+ per gram of chitosan and a final pH value of 2.06 were reached 140 hours after the addition of nitric acid. It is notable that the hydrogen ion consumption of 8.9 mmole H^+/g chitosan is consistent with the hydrogen ion consumption data presented in Figure 2 (b).

3.2. Optimum pH and Hydrogen Ion Consumption

Hydrogen ions are needed to replace the cadmium ions adsorbed on the amine groups of the chitosan beads. Therefore, different amounts of hydrogen ion were added into the vessel to determine the effect of hydrogen ion consumption on the percentage of cadmium desorbed and the final pH value of the cadmium solution obtained in the single stage desorption process. The percentage of cadmium desorbed is presented in Figure 4 as a function of desorption time for different amounts of hydrogen ion initially charged to the vessel. Specifically, after cadmium adsorption, different amounts of 0.1 N nitric acid (22.2 mL, 50 mL and 65 mL respectively) were dosed into the reactor. At a final pH

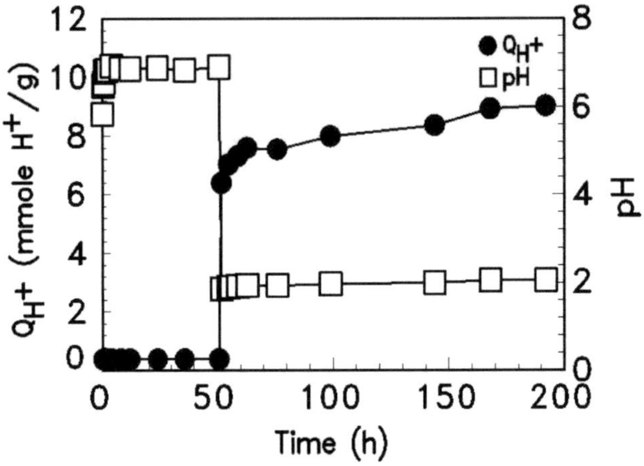

Figure 3. The hydrogen ion consumption and pH versus time for cadmium-free water saturation experiment

value of 2.0, 94 % of cadmium desorption was achieved and 8.3 mmole H^+ per gram of beads were consumed. It is interesting to note that even at a final pH of 4.7, 80 % of cadmium desorption was accomplished.

Stage	Time (h)	pH	D (%)
A-1	48	7.00	
D-1	67	6.80	7.01
D-2	78	6.64	27.42
D-3	90	5.80	39.75
D-4	101	5.48	39.33
D-5	119	4.36	84.66
D-6	141	3.00	95.72

Table 4. Table 4. The effect of pH adjustment on the % of cadmium desorbed (D) for the multiple stage desorption experiment

The multiple stage desorption experiment was carried out by a series addition of nitric acid solution into the spinning basket reactor. The effect of pH on the percentage of cadmium desorbed is given in Table 4. After 6 stages of desorption in series, 95.7 % of the cadmium desorption was achieved at a final pH of 3.0.

Even at an equilibrium pH of 4.36, 84.6 % of cadmium desorption was reached, consistent with the single stage desorption process Figure 4

Previous researchers used shake flask experiments to study the effect of pH on heavy metal ion adsorption capacity. Eric and Roux (1992) used the shake flask experiment to study the influence of pH on the heavy metal ion binding onto a fungus-derived bio-sorbent. Inoue et al. [10]

also used the shake flask experiment to evaluate the effect of the hydrochloric acid concentration on the adsorption of platinum group metal ions onto chemically modified chitosan.

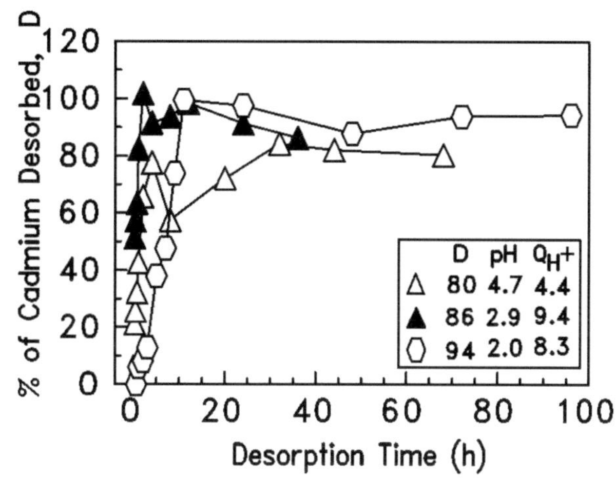

Figure 4. The percentage of cadmium desorbed versus desorption time for different amounts of hydrogen ion initially charged to the vessel

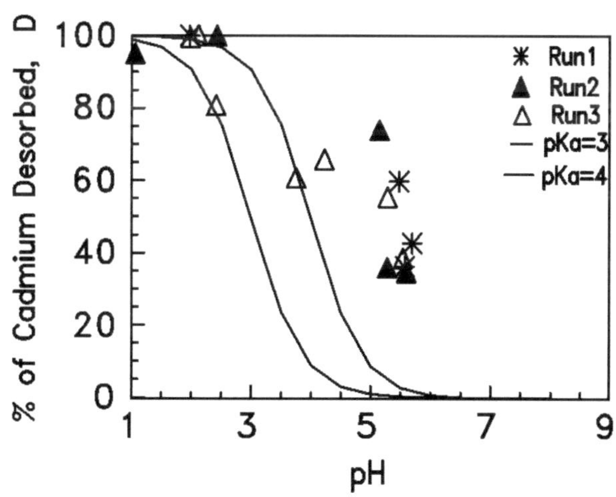

Figure 5. The experimental and predicted percentages of cadmium desorbed versus equilibrium pH for shake flask experiments

Shake flask desorption experiments were also performed in this present study and compared to the results from the spinning basket reactor experiments and to previous works. Data for the percentage of cadmium desorbed is plotted as a function of the equilibrium pH in Figure 5. At an equilibrium pH of 2.1, 100 % cadmium desorption was achieved. Increasing the equilibrium pH decreased the percentage of cadmium desorbed according to an S-shaped profile. This S-shaped profile was also observed by Schultz et al. [11] and Aldor et al. [12]. Jha et al. [6] also found that 88 % of cadmium desorption was obtained 24 hours after addition of 0.01 N hydrochloric acid to 100 mg of cadmium-loaded chitosan. Also, 80 % cadmium desorption occurred at pH 3.0, a similar result as the present study. Hayes and Leckie [13] presented an S- shaped profile for Cd^{+2} and Pb^{+2} adsorption on goethite, which has a complimentary profile to the present desorption study.

Results of the percentage of cadmium desorbed from the shake flask desorption experiment were consistent with the spinning basket desorption experiment. However, the results for hydrogen ion consumption were quite different. The inconsistent hydrogen ion consumption may be due to the sensitivity of the pH measurement. A small change in pH near 2 resulted in a significant difference in hydrogen ion consumption. Also the modes of mixing between the spinning basket reactor experiment and the shake flask experiment were different. The mixing in the spinning basket reactor was more uniform and resulted in more repeatable and reliable pH measurements. Therefore, the measurements for H^+ adsorption on the chitosan beads for the shake flask experiments at low pH need to be interpreted with caution.

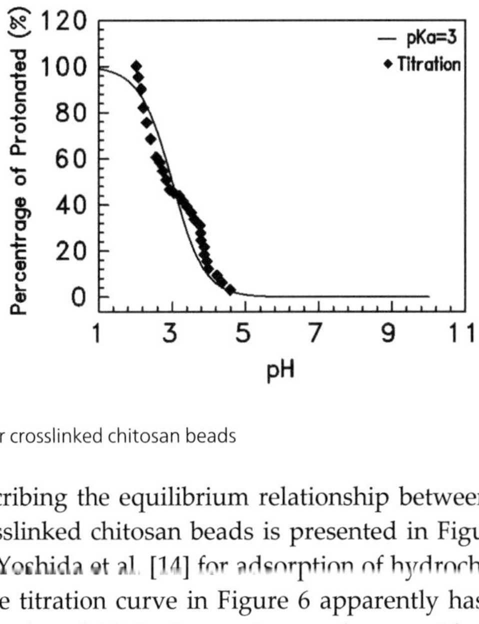

Figure 6. The titration curve for crosslinked chitosan beads

The titration curve describing the equilibrium relationship between the pH and hydrogen ion capacity on the crosslinked chitosan beads is presented in Figure 6. A similar titration curve was observed by Yoshida et al. [14] for adsorption of hydrochloric acid on poly(ethylene imine) chitosan. The titration curve in Figure 6 apparently has two equivalent points. For the first equivalent point of 4.0, hydrogen ions exchange with the imino groups (=NH) on the chitosan beads within the crosslinked outer shell. The second equivalent point at pH of 2.5 shows that hydrogen ions penetrate deeper to the un-crosslinked core of bead to exchange with the free amine groups ($-NH_2$).

3.3. Exchange Between Cadmium Ions and Hydrogen Ions

The proposed competitive ion exchange scheme for cadmium ions and hydrogen ions with nitrogen atoms on crosslinked chitosan is presented in Figure 7. The addition of hydrogen ions displaces the adsorbed cadmium ions. A Langmuir-Freundlich equation for multiple components is developed below to model the competition between the hydrogen ions and the adsorbed cadmium ions on the crosslinked chitosan beads.

Model Development. Three major assumptions are made for the equilibrium model. First, the adsorption of cadmium ions on the crosslinked chitosan beads may follow the chelation binding mechanism validated by Inoue et al. [15], which shows that divalent cadmium ions adsorb onto amine groups of chitosan to form metal-chelate complexes with composition of 1 mole of cadmium to 2 mole of glucosamine unit. Second, the Cd^{+2} ions chelate only with imino (-CH=N-) groups in the outer shell of the crosslinked chitosan. In Chapter 3, it was shown that the crosslinked chitosan bead contains a crosslinked outer shell where all amine groups are derivativized to imino

(-CH=N-) crosslinks, and an inner core of uncrosslinked chitosan containing free amine (-NH_2) groups. In the present desorption experiments, the maximum adsorption capacity was around 15 mg Cd^{+2}/g chitosan. Thus, the value for r_M from previous study [8] is 1.32 mm, which is larger than r_c value of 1.29 mm for the crosslinked chitosan beads used in the desorption experiments. Consequently the adsorbed cadmium is only localized in the cross-linked outer shell of the chitosan beads where only imino chelation sites are present. Therefore, the chelation reaction is expressed as

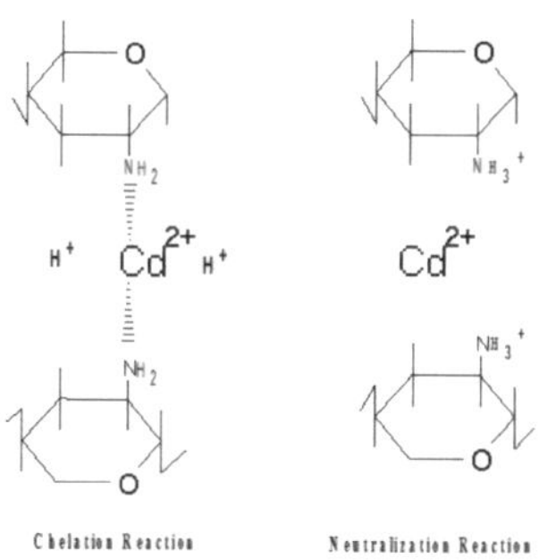

Figure 7. The competitive ion exchange scheme

$$2 - C = NH + Cd^{+2} \overset{K_{Cd}}{\Leftrightarrow} 2\ \text{-C=NH} \bullet Cd^{+2} \tag{6}$$

The third assumption is that the equilibrium constants for the adsorption of hydrogen ions on the chitosan beads may follow the acid/base neutralization reactions of the form

$$- C = NH + H^{+} \overset{K_a'}{\Leftrightarrow} \text{-C=}NH_2^{+} \tag{7}$$

$$\text{or} \quad \text{-C-NH}_2 + H^+ \overset{K_a'}{\Leftrightarrow} \text{-C-NH}_3^+ \tag{8}$$

In other words, the equilibrium constant K_a' is assumed to be the same for either imino or amine groups. Consequently, Figure 6 is assumed to have only one apparent pK_a value at pH 3.0. From equations (6) and (7), K_{cd} and K_a' are defined as

$$K_{Cd} = \frac{[\text{-C=NH} \bullet Cd^{+2}]}{[\text{-C=NH}][Cd^{+2}]^{1/2}} \tag{9}$$

$$K_a' = \frac{[\text{-C=NH}_2^+]}{[\text{-C=NH}][H^+]} \tag{10}$$

If the overall active number of adsorption sites on the crosslinked chitosan bead is conserved, then the equilibrium constants K_{cd} and K_a'can be expressed as

$$K_{Cd} = \frac{2\, q_{Cd}}{(Q_M - 2\, q_{Cd} - q_H)\, C_e^{\,1/2}} \tag{11}$$

$$K_a' = \frac{q_H}{(Q_M - 2\, q_{Cd} - q_H)\, C_{ae}} \tag{12}$$

where C_e is the equilibrium cadmium ion concentration in the vessel (mole /L), C_{ae} is the equilibrium hydrogen ion concentration in the cadmium solution (mole/L), K_a' is the equilibrium constant $(\text{mole/L})^{-1}$ for the neutralization reaction, K_{cd} is the equilibrium constant $(\text{mole/L})^{-1/2}$ for chelation, Q_M is the theoretical maximum capacity of chitosan (6.2 mmole active sites/g chitosan), q_{Cd} is the equilibrium cadmium ion capacity of the crosslinked chitosan beads (mmole Cd^{+2}/g chitosan), and q_H is the equilibrium hydrogen ion capacity of the crosslinked chitosan beads (mmole H^+/g chitosan).

By solving for q_{Cd} and q_H from equations (10) and (11), the following Langmuir-Freundlich equations were derived, based on the binary mixture model presented by Ruthven [16]:

$$q_{Cd} = \frac{K_{Cd}\, C_e^{\,1/2}\, Q_M / 2}{1 + K_a'\, C_{ae} + K_{Cd}\, C_e^{\,1/2}} \tag{13}$$

$$q_H = \frac{K_a'\, C_{ae}\, Q_M}{1 + K_a'\, C_{ae} + K_{Cd}\, C_e^{\,1/2}} \tag{14}$$

The sum of q_{Cd} and q_H is expressed as q, given by

$$q = q_{Cd} + q_H = \frac{K_{Cd}\, C_e^{\ 1/2}\, Q_M / 2 \ + \ K_a{}'\, C_{ae}\, Q_M}{1 + K_a{}'\, C_{ae} + K_{Cd}\, C_e^{\ 1/2}} \tag{15}$$

In equations (12) and (14), $Q_M/2$ is equal to 3.1 mmole Cd^{+2}/g chitosan.

Recall from equation (3) that the percentage of cadmium desorbed (D) is given by

$$D\,(\%) = \left(1 - \frac{q_{Cd}}{q_{Cd,f}}\right) \bullet 100\,\% \tag{16}$$

where $q_{Cd,f}$ is the final cadmium adsorption capacity before the desorption process is initiated (mmole Cd^{+2}/g chitosan). Therefore, the percentage of the cadmium desorbed into the cadmium solution during the desorption process is expressed as

$$D\,(\%) = \left(1 - \frac{K_{Cd}\, C_e^{\ 1/2}\, Q_M / 2}{1 + K_{Cd}\, C_e^{\ 1/2} + K_a{}'\, C_{ae}} \ \frac{1}{q_{Cd,f}}\right) \bullet 100\,\% \tag{17}$$

Estimation of Equilibrium Constants. The pK_a value of the crosslinked chitosan beads was obtained from the pH value at which 50 % of the active sites were protonated. The pK_a value estimated from Figure 6 was 3.0. In other words, the pK_b value of the conjugated base was 11.0 (Snoeyink and Jenkins, 1982 [17]). In this study $K_a{}'$is defined as the inverse of K_a. Based on the relationship between pK_a and K_a (pK_a = - log K_a), the value of $K_a{}'$was equal to 10^{pKa}. Therefore, the $K_a{}'$value was 1000 $(mole/L)^{-1}$. Muzzarelli [18] found that the pK_a value for chitosan was 6.3. Similarly, the pK_b for the base $-C-NH_2$ was 7.7. The pK_b differences may suggest that the crosslinked imino group (=NH) on chitosan is a weaker base than the uncrosslinked amine group ($-NH_2$). Both the titration curve for adsorption of hydrochloric acid onto poly(ethylene imine) chitosan beads and the derived pK_a value of 4.0 by Yoshida et al. [14] are similar with the results of this present work.

Once $K_a{}'$was obtained, K_{cd} was estimated. The $q_{Cd,f}$ values at C_e around 200 mg Cd^{+2}/L (1.79 mmole Cd^{+2}/L) and pH at 6.5 to 7 were between 0.13 and 0.17 mmole Cd^{+2}/g chitosan. After substitution of $K_a{}'$, C_{ae}, C_e, Q_M, and $q_{Cd,f}$ values into equation (15), K_{cd} was determined when the minimum sum of squares between the data points given in Figure 4.5 for the percentage of cadmium desorbed (D) and the predicted values from equation (15) was achieved. The K_{cd} value was estimated to be 0.57 $(mole/L)^{-1/2}$. This chelation reaction constant was also compared to the stability constant data for metal complexes on organic ligands (IUPAC, 1979) which showed that the binding constant of cadmium on pyridine (imine ligand) and sterotonin (amine ligand) were $10^{\ 1.36}$ and $10^{\ 3.6}$ $(mole/L)^{-1/2}$ respectively. Therefore, the value for estimated K_{cd} indicated that the cadmium did not bind very strongly to imino groups on chitosan. This low binding constant facilitated the desorption process as the hydrogen ions easily displaced the cadmium ions to accomplish the neutralization reaction. However, hydrogen ions were still required for desorption because water alone could not desorb the bound cadmium.

Another approach for the calculation of K_{cd} was considered by using the low cadmium concentration adsorption isotherm data for 2.5 wt % crosslinked chitosan beads in previous study [8]. In the previous study, q_{Cd} is equal to 30.66 mg Cd^{+2}/g chitosan (0.27 mmole/g chitosan) at an equilibrium cadmium concentration of 234.4 mg Cd^{+2}/L (2.09 mmole Cd^{+2}/L). According to equation (12), when the pH is greater than or equal to 7.0, the product of the K_a' C_{ae} will be much smaller than the product of $K_{Cd}C_e^{\ 1/2}$. The value of K_{Cd} was determined by setting C_e, Q_M and q_{Cd} equal to 2.09 mmole/L, 6.2 mmole/g, and 0.27 mmole/g respectively in equation (12). The K_{cd} value estimated at pH 7.0 was 2.12 $(mole/L)^{-1/2}$ which is higher than the K_{cd} value of 0.57 $(mole/L)^{-1/2}$ estimated by equation (15). However, the values of K_{Cd} estimated by both methods are in general much smaller than the value of K_a'. The value of K_{cd} is not sensitive to the prediction of the percentage of cadmium desorbed by Langmuir-Freundlich equation, since the K_a'value was three orders of magnitude bigger than the value of K_{cd}. Thus, the K_a'value (pK_a value) determines how adsorbed cadmium ions are displaced by hydrogen ions. In other words, the value of the $K_a'$is the limiting parameter for the predicting desorption profile.

At pH ranging from 1 to 8, the percentage of cadmium desorbed was predicted using equation (15). The experimental and the predicted percentages of cadmium desorbed are plotted in Figure 5 as a function of pH. In Figure 5, the majority of the adsorbed cadmium ions were desorbed at pH ranging from 3.0 to 5.5. As mentioned previously in Figure 6, the titration curve has two equivalent points at pH of 2.5 and 3.9. Since the experimentally occurred percentage of cadmium desorbed was significant even at pH values above 4.0, the first equivalent point at pH 3.9 may determine the actual pK_a required for desorption. Therefore, for comparison, the predicted desorption profile assuming pK_a equal to 3.9 is also provided in Figure 5.

The experimental data for three repeat runs is scattered in the pH range from 4.5 to 6.0, as shown in Figure 5. Equation (15) does not accurately predict the behavior of desorption in this pH range. At pH ≥5.0, the desorption process may be more complex than the model suggests. The ion exchange behavior can not fully predict the chelation reaction.

4. Summary and Conclusions

Chitosan is nature's most abundant biopolymer next to cellulose. It is well known that chitosan is a selective adsorbent for heavy metal ions. However, the chitosan raw material needs to be modified for use in low pH environments. To address this need, chemically modified porous chitosan beads were synthesized.

In order to fabricate porous chitosan beads, chitosan in acetic acid solution was cast into spherical gel beads of 3 mm diameter and precipitated to a gel in 2 M NaOH solution.

http://dx.doi.org/10.5772/50142

Chemical modifications of chitosan were considered. First, the linear chitosan chains within gel beads were heterogeneously crosslinked by glutaric dialdehyde solution. The chitosan gel beads were then freeze dried to form porous beads.

The desorption of cadmium from cadmium-adsorbed chitosan beads with dilute nitric acid was tested to evaluate the feasibility of recovering the cadmium and regenerating the adsorbent. Hydrogen ions were needed to displace the cadmium ions adsorbed on the chitosan beads. Two types of desorption experiments, single stage and multiple stage, were performed using a Carberry spinning basket reactor. In the experiments, 0.5 g of chitosan beads were packed into the hollow impeller basket assembly and contacted with 200 mL of 200 mg /L cadmium ion solution at 150 rpm and 25 C until the adsorption equilibrium was achieved. After adsorption, different doses of 0.1 N HNO_3 solution were added to the vessel to initiate the desorption process. The cadmium concentration increased sharply during the first 12 hours of desorption following the addition of nitric acid, and then leveled off. At a final pH value of 2.0, 94 % of cadmium desorption was achieved, and 8.3 mmole H^+ per gram of beads was adsorbed to displace the bound cadmium. Equilibrium shake flask experiments were carried out to determine the equilibrium hydrogen and cadmium loading on the chitosan beads at different pH levels. Decreasing the equilibrium pH increased the percentage of cadmium desorbed according to an S-shaped profile, consistent with the ion-exchange mechanism. A Langmuir-Freundlich model proposed that the desorption process is accomplished by displacing adsorbed cadmium ions with hydrogen ions. Based on the high efficiency of the desorption treatment, cadmium recovery and adsorbent regeneration is feasible.

In summary, the adsorbent can be regenerated by dilute acid treatment, and 100 % cadmium recovery from the chitosan beads is feasible at pH less than 3.0. Two recommendations for future research are suggested:

1. Many industrial waste water matrices contain several heavy metal ions. Therefore, determine the multi-component heavy metal ion adsorption capacity on the chitosan beads.

2. Modify the bead synthesis process. The freeze drying process does not improve the adsorption capacity. Therefore, consider alternative drying methods that do not reduce the adsorption capacity and are less expensive than freeze drying.

5. Nomenclature

$C(t)$	cadmium ion concentration in the vessel at different adsorption or desorption times, mg Cd/L
C_a	concentration of HNO_3 added into the spinning basket reactor before desorption, mole H^+/L
C_{ae}	equilibrium hydrogen ion concentration in the cadmium solution, mole/L
C_e	equilibrium cadmium ion concentration in the vessel, mole /L
C_o	initial concentration of Cd^{+2}, mg Cd^{+2}/L
C_f	final concentration of Cd^{+2} at equilibration, mg Cd^{+2}/L

C_{min}	lowest cadmium concentration along the adsorption or desorption process, mg Cd/L
$C_{min,i}$	lowest cadmium concentration along the adsorption or desorption process at a specific desorption stage, mg Cd/L
$C_{0,i}$	initial cadmium concentration in the spinning basket reactor at specific desorption stage " i " , mg Cd/L
D	percentage of cadmium desorbed at a given desorption time, %
D_{Ae}	effective diffusion coefficient of glutaric dialdehyde within the crosslinked zone of the gel bead, cm2/sec
K	measure of the adsorption capacity or binding strength
Ka'	equilibrium constant for the neutralization reaction, $(mole/L)^{-1}$
K_{cd}	equilibrium constant for chelation, $(mole/L)^{-1/2}$
$M_{w,Cd+2}$	molecular weight of the cadmium ions, g/mole
m_b	mass of chitosan beads in the spinning basket reactor impeller, g
n	number of desorption stage
1/n	measure of adsorption intensity
pH_t pH	value at different adsorption or desorption times
Q	cadmium adsorption capacity on the chitosan beads, mg Cd/g-chitosan
Q_f	final cadmium adsorption capacity on the chitosan beads, mg Cd^{+2}/g chitosan
$Q(H^+)$	accumulated hydrogen ion adsorption capacity in the present stage, mg Cd/g-chitosan
q_{Cd}	equilibrium cadmium ion capacity of the crosslinked chitosan beads, mmole Cd^{+2}/g chitosan
Q_M	theoretical maximum capacity of chitosan, 6.2 mmole active sites/g chitosan
q_H	equilibrium hydrogen ion capacity of the crosslinked chitosan beads, mmole H+/g chitosan
R	radius of the gel bead, cm
r	radial position within the gel bead, cm
r_C	radial position within the gel bead defining the boundary between the outer crosslinked zone and the inner unreacted core zone, cm
r_M	cadmium un-saturation zone, cm
t	time of crosslinking
V	current cadmium solution volume loaded in the spinning basket reactor vessel, L
V_a	acid volume added into the spinning basket reactor, L
V_t	volume of solution in the spinning basket reactor at a given time, L
$X_{Bgroups}$	weight fraction of chitosan in the gel bead, g chitosan/g of gel bead extent of crosslinking, moles of crosslink/total moles of $-NH_2$
X_T	moles of glutaric dialdehyde consumed by the gel bead/total moles of $-NH_2$ groups within the gel bead
X_R	moles of glutaric dialdehyde crosslinked per total mole of $-NH_2$ groups within the gel bead

Y_B	total moles of amine groups/g of chitosan
β	moles of glutaric dialdehyde consumed to form a crosslink/moles of $-NH_2$ crosslinked
ρ_b	overall density of the gel bead, g/cm^3
υ•	chelation coordination number for cadmium, 2 moles active sites/mole Cd^{+2}
ΔQi (H^+)	hydrogen ion adsorption capacity for chitosan beads at the ith desorption stage mg H/g-chitosan

Author details

Tzu-Yang Hsien[1*] and Yu-Ling Liu[2]

*Address all correspondence to: tyhsien@cute.edu.tw

1 General Education Center, China University of Technology, Taipei

2 Teacher Education Center. Ming Chuan University, Taoyuan

References

[1] Rozzlle, L. T. (1987). *J.Am. Water Works Assn.*, 10, 53-59.

[2] Yoo, R. S. (1987). *Am. Water Works Assn.*, 10, 34-38.

[3] Rorrer, G. L., Hsien, T. Y., & Way, J. D. (1993). Synthesis of Porous-Magnetic Chitosan Beads for Removal of Cadmium Ions from Waste Water. *I & EC Research*, 32, 2170-2178.

[4] Muzzarelli, R. A. A., & Rocchetti, R. (1974). The Use of Chitosan Columns For the Removal Of Mercury From Eaters. *J. Chromatog.*, 96, 115-121.

[5] Randall, J. M., Randall, V. G., Mc Donald, G. M., Young, R. N., & Masri, M. S. (1979). Removal of Trace Quantities of Nickel from Solution. *Journal of Applied Polymer Science*, 23, 727-732.

[6] Jha, I. N., Iyengar, L., & Rao, A. V. S. (1988). Removal of Cadmium Using Chitosan. *J. Environs. Eng.*, 114, 962-974.

[7] Coughlin, R. W., Deshaies, M. R., & Davis, E. M. (1990). Chitosan in Crab Shell Wastes Purifies Electroplating Wastewater. *Environmental Progress*, 9(1), 35-39.

[8] Hsien, T. Y., & Rorrer, G. L. (1995). Effects of Acylation and Crosslinking on the Material Properties and Cadmium Ion Adsorption Capacity of Porous Chitosan Beads. Seperation Science and Technology - , 12(30), 2455-2475.

[9] Carberry, J. J. (1976). Chemical and Catalytic Reaction Engineering. *Chemical Engineering Series*, McGraw-Hill, Inc.

[10] Inoue, K., Baba, Y., Yoshizuka, K., Noguchi, H., & Yoshizaki, M. (1988). Selectivity Series in the Adsorption of Metal Ions on a Resin. *Crosslinking Copper (II) Complexed Chitosan, Chem. Lett.*, 1281-1284.

[11] Schultz, M. F. (1987). Adsorption and Desorption of Metals On Ferrihydrite: Reversivility of Reaction and Sorption Properties of the Regenerated Solid. Environ. *Sci. Technol*, 21, 863-869.

[12] Aldor, I., Fourest, E., & Volesky, B. (1995). Desorption of Cadmium from Algal Biosorbent. *The Canadian Journal Of Chemical Engineering*, 73, 516-522.

[13] Hayes, K. F., & Leckie, J. O. (1987). Modeling Ionic Strength Effects on Cation Adsorption at Hydrous Oxide/Solution Interfaces. *Journal of Colloid and Interface Science*, 115, 564-572.

[14] Yoshida, H., Kishimoto, N., & Kataoka, T. (1994). Adsorption of Strong Acid on Polyaminated Highly Porous Chitosan: Equilibris. *Ind. Eng. Chem. Re.*, 33, 854-859.

[15] Inoue, K., Yamaguchi, T., Iwaski, M., Ohto, K., & Yoshizuka, K. (1995). Adsorption of Some Platinum Group Metals on Some Complexane Types of Chemically Modified Chitosan. *Separation Science and Techno.*, 30(12), 2477-2489.

[16] Ruthven, D. M. (1984). Langmuir-Freundlich Equations. *Principles of Adsorption & Adsorption Processes*, John Wiley & Sons, Inc.

[17] Snoeyink, V., & Jenkins, D. (1982). *Water Chemistry*, 86-120.

[18] Muzzarelli, R. A. A. (1977). *Chitin, Pergamon Press.*

Section 4

Solar Desalination

Chapter 9

Experimental Study on a Compound Parabolic Concentrator Tubular Solar Still Tied with Pyramid Solar Still

T. Arunkumar, K. Vinothkumar, Amimul Ahsan,
R. Jayaprakash and Sanjay Kumar

Additional information is available at the end of the chapter

http://dx.doi.org/10.5772/49995

1. Introduction

Water is a nature's gift and it plays a key role in the development of an economy and in turn for the welfare of a nation. Non-availability of drinking water is one of the major problem faced by both the under developed and developing countries all over the world. Around 97% of the water in the world is in the ocean, approximately 2% of the water in the world is at present stored as ice in polar region, and 1% is fresh water available for the need of the plants, animals and human life [1]. Researchers have been carried out in this method by Nijmeh, et al., [2], they have been investigated the regenerative, conventional and double-glass-cover cooling solar still theoretically and experimentally. Several system parameters were also investigated with respect to their effect on the productivity, namely, water with and without dye in the lower basin, basin heat loss coefficient, and mass of water in the basins and mass flow rate into the double-glass cover. Thermal performances of a solar still coupled with flat plate heater along with an evaporator-condenser have been analyzed by René Tchinda, et al., [3]. They reported that the theoretical solar still productivity is in reasonably good agreement with the experimental distillation yields. Thermal performances of a regenerative active solar distillation system working under the thermosyphon mode of operation have been studied by Singh and Tiwari for Indian climatic condition. It is concluded that (i) there is a significant improvement in overall performance due to water flow over the glass cover and (ii) the hot water available due to the regenerative effect does not enhance the output. They derived expressions for water and glass temperatures, hourly yield and instantaneous efficiency for both passive and active solar distillation systems [4]. Chouchi et al., [5] have designed and built a small solar desalination unit equipped with a parabolic concentrator. The results show that, the maximum efficiency corresponds to the maximum solar light-ning obtained towards 14:00. At that hour, the boiler was nearly in a horizontal position, which maximizes the offered heat transfer surface. Thermal analyses of a concentrator assisted regenerative solar distillation unit in forced circulation mode were studied by Kumar and Sinha [6]. It is concluded that the yield of the concentrator assisted regenerative solar still is much higher than any other passive/active regenerative or non-regenerative solar distillation system and the overall thermal efficiency increases with an increase in the flow rate of the flowing cold water over the glass cover. Thermal evaluations of concentrator assisted solar distillation system have been studied by Sinha and Tiwari [7]. It was observed that the instantaneous and overall efficiency of the concentrator assisted solar still is significantly improved compared to a collector cum distillation unit due to reduced heat loss in the concentrator. Tube-type solar still integrated by a conventional still and a water distribution network suitable to the concept of desert plantation was studied by Murase, et al., [8]. Experimental data measured using infrared lamps which showed the effectiveness of the method for productivity, the design of the basin tray and thermal efficiency. Tiwari and Kumar [9] have experimentally studied the tubular solar still. The still consists of a rectangular black metallic tray placed at the diametric plane of a cylindrical glass tube. It was concluded that the daily yield of distillate in the tubular solar still is higher than that of the conventional solar still. The purity of the product in the tubular solar still is higher than that of the conventional one, and could be used for chemical laboratories, etc.

A new heat and mass transfer for tubular solar still was studied by Islam and Fukuhara [10]. The heat and mass transfer coefficients were expressed as functions of the temperature difference between the saline water and the cover. A quasi steady heat and mass transfer tubular solar still taking an account of humid air properties inside the still was analyzed by Islam and Fukuhara [11]. It was found from the production experiment that the analytical solutions derived from the present model could reproduce the experimental results on the saline water temperature, the humid air temperature, the cover temperature and production and condensation fluxes. Ahsan et al., [12] has experimentally studied the evaporation, condensation and production of a tubular solar still. They found that the relative humid of the humid air was definitely not saturated and the hourly evaporation, condensation and production fluxes were proportional to the humid air temperature and relative humidity.

This paper covers an experimental study on a compound parabolic concentrator tubular solar still coupled pyramid solar still with and without top cover cooling has been investigated.

2. Fabrication Details

The inner and outer tubes are positioned with a 5 mm gap for flowing cold water to cool the outer surface of the inner glass tube. A circular basin of dimension 2m length and a diameter 0.035 m was designed and coated with black paint using a spray technique. Pyramid solar still of area 1 m x 1 m is designed. The bottom of the still is insulted using saw dust. The solar still is insulated with saw dust reduces the cost of fabrication. Consequently, the cost for fresh water production is less. In the view of eco-friendly material, saw dust would be a good alternative for glass wool. The pyramid solar still is coupled with a non-tracking CPC with help of insulated pipes. The top cover is cooled by flowing cold water at a constant

http://dx.doi.org/10.5772/49995

flow rate of 10ml/min. It is adjusted by using a pressure head. It is adjusted for maintaining constant water level in the water storage tank initially during the experiment. A graduated measuring jar is used to measure the flow rate. The process is repeated many times until steady cold water flow in between the tubular cover. The following parameters were measured every fifteen minutes of interval. Water temperature (T_w), interior humid air temperature (T_a), ambient temperature (T_{amb}), outer cover of the tubular temperature (T_{oc}), total radiation (I_{diff}) and direct solar radiation (I_{dir}), and distillate yield. The radiation is measured by a Precision Pyranometer and Pyrheliometer.

3. Experimental Setup

The experimental setup of the system is shown in Figs. 1-3. The distilled yield extracted from both CPC tubular solar still and pyramid solar still. The pyramid solar still is directly coupled with compound parabolic concentrator (CPC) through an insulated pipe. The cold water from the water tank is passed to cool the tubular cover of the still through inlet. The heat energy gained from the top cover cooling process is extracted through outlet and stored in the basin of four sloped solar still. The basin water temperature is raised to a maximum level within a short period, the operating temperature of the still becomes higher and distillation has been started. As well as the radiation falls on the surface of the pyramid solar still which keeps the temperature at a constant level than it reduces through convection. The condensate yield is started to increase due to the temperature difference between water in the basin and top cover of the pyramid still top cover temperature is decreased by cold water flow over it. Thus the temperature difference is wider and produces a distillate yield to a larger quantity.

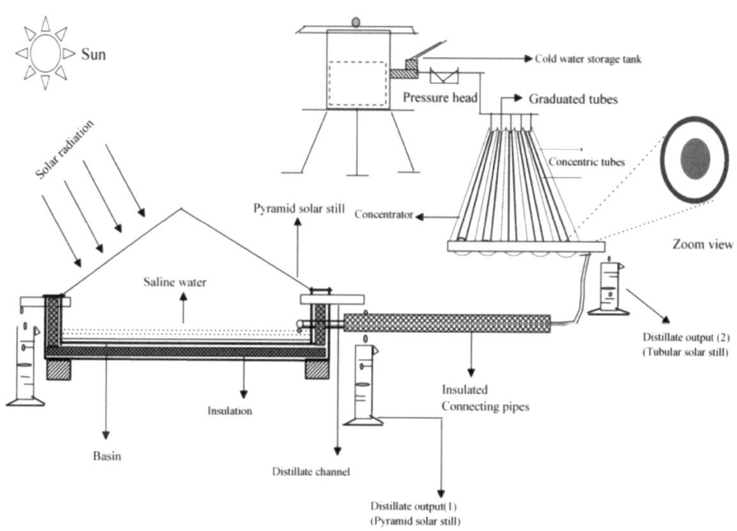

Figure 1. Cross sectional view of compound parabolic concnetrator tubular solar still coupled with pyramid solar still.

Figure 2. Pictorial view of compound parabolic concnetrator tubular solar still coupled with pyramid solar still.

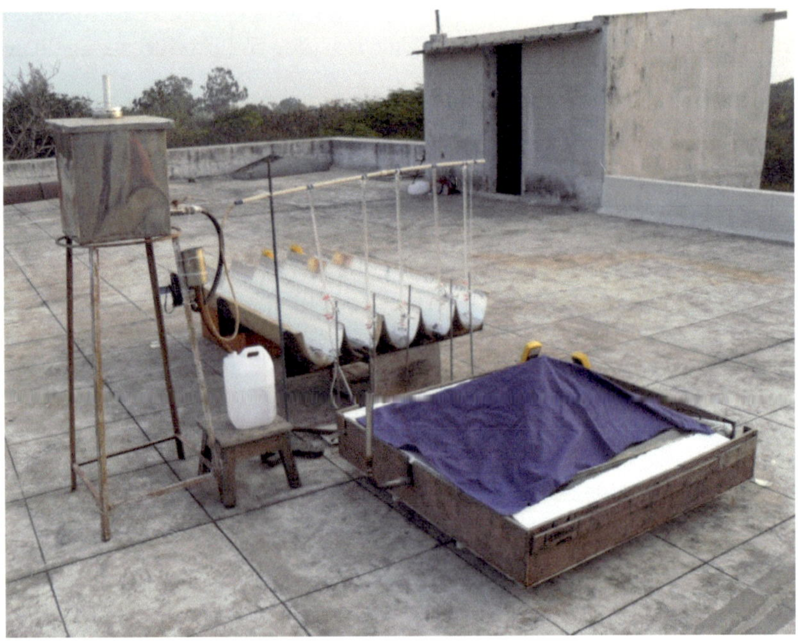

Figure 3. Pictorial view top cover closed pyramid solar still coupled with compound parabolic concentrator tubular solar still.

http://dx.doi.org/10.5772/49995

4. Results and Discussion

Total radiation with respect to time is shown in Fig. 4. Hourly variation of solar radiation is in the range of 520-1036 W/m^2 for CPC-CCBTSS –during top cooling at the tubular solar still and the average solar intensity is 791.72 W/m^2. Similarly the radiation measured as 495-1060 W/m^2 for CPC-CCBTSS-Pyramid solar still top cover without cooling and the average radiation is 793.42 W/m^2. The radiation measured in the range of 579-1050 W/m^2 during the study of effect of top cover cooling in pyramid solar still and the average radiation is 790.27 W/m^2.

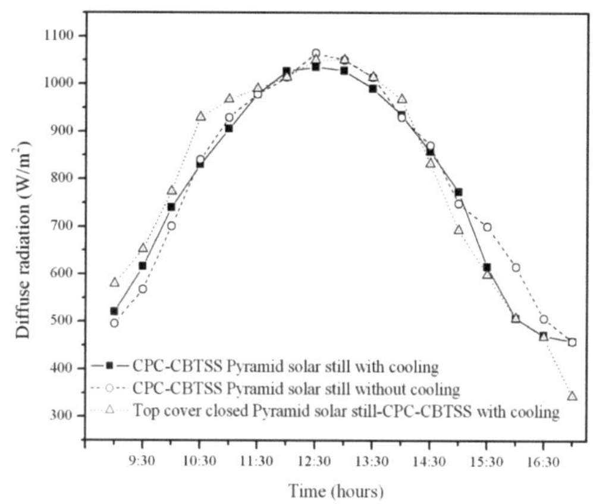

Figure 4. Hourly variation of diffuse solar radiation with respect to time.

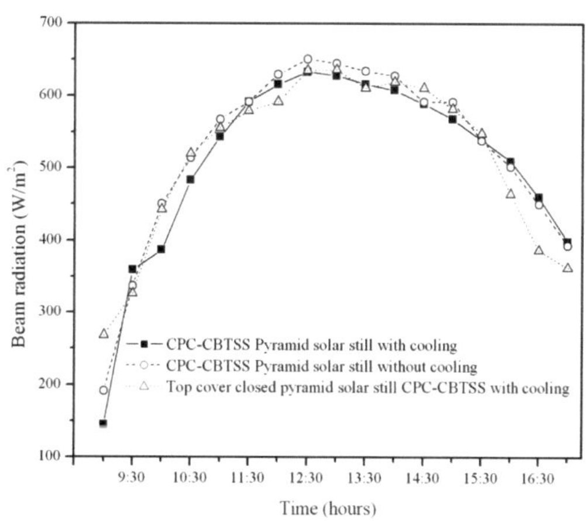

Figure 5. Hourly variation of beam solar radiation with respect to time.

Direct radiation with respect to time is shown in Fig. 5. Hourly variation of direct radiation is in the range of 244-735 W/m^2 for CPC-CCBTSS with Pyramid solar still's top cover cooling process and the average solar intensity is 610.20 W/m^2. Similarly 291-751 W/m^2 for CPC-

CCBTSS-Pyramid solar still top cover without cooling and the average radiation is 623 W/m², and 268-739 W/m² for top cover closed pyramid solar still with CPC-CCBTSS and the average radiation is 613.90 W/m². Hourly variation of ambient temperature is shown in Fig. 6. The recorded ambient temperature is in the range of 29°C to 36°C for cooling. All the analyses are carried out in almost same atmospheric effect same during the study and it is more compatible for comparison. Similarly, 28°C to 35°C for CPC-CCBTSS pyramid solar still top cover cooling and 28°C to 36°C for top cover closed CPC-CCBTSS with cooling.

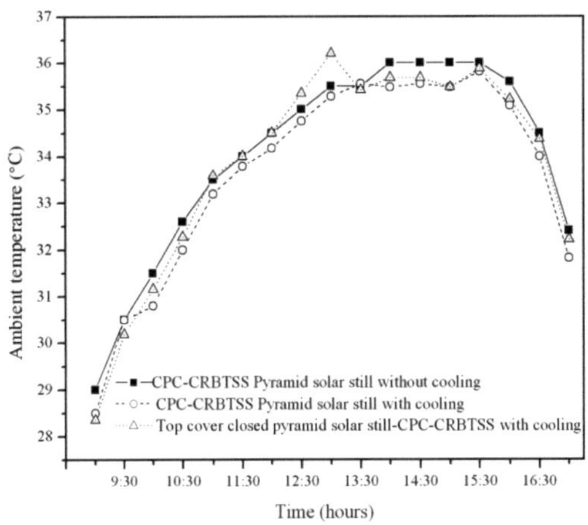

Figure 6. Hourly variation of ambient temperature with respect to time.

Fig. 7 shows the variation of water temperature, air temperature, and outer cover temperature with respect to time in the pyramid solar still. The maximum rise in water temperature is observed as 68°C. Similarly the maximum air temperature inside the pyramid still as 60°C and outer cover temperature is 43°C. Fig. 8 shows the variation of water temperature, air temperature, and outer cover temperature with respect to time for tubular solar still with circular basin. The maximum rise in water temperature is observed as 95°C, the maximum air temperature as 80°C and maximum outer cover temperature as 54°C are obtained from this study. Fig. 9 shows the variation of water temperature, air temperature, and outer cover temperature with respect to time for pyramid solar still top cover cooling. The maximum rise in water temperature is observed as 77°C. Similarly the maximum air temperature of 69°C and the outer cover temperature of 42°C are obtained. Fig.10 shows the variation of water temperature, air temperature, and outer cover temperature with respect to time for tubular solar still circular basin with water cooling. Similarly, the maximum rise in water temperature, air temperature and outer cover temperature are measured as 77°C, 67°C and 50°C respectively. Fig. 11 shows the variation of water temperature, air temperature, and outer cover temperature with respect to time for top cover closed pyramid solar still coupled tubular solar still. The maximum rise in water temperature is observed as 60°C. Similarly the maximum air temperature of 49°C is measured and the maximum outer cover temperature as 37°C is measured.

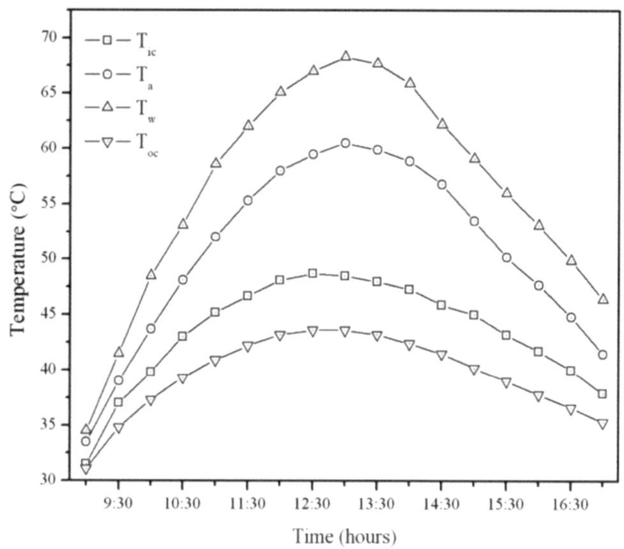

Figure 7. Hourly variation of temperatures with respect to time in pyramid solar still.

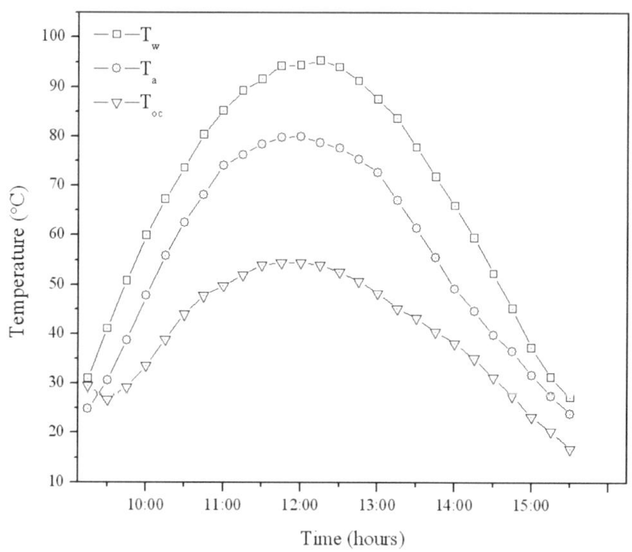

Figure 8. Hourly variation of temperature in tubular solar still in CPC assembly.

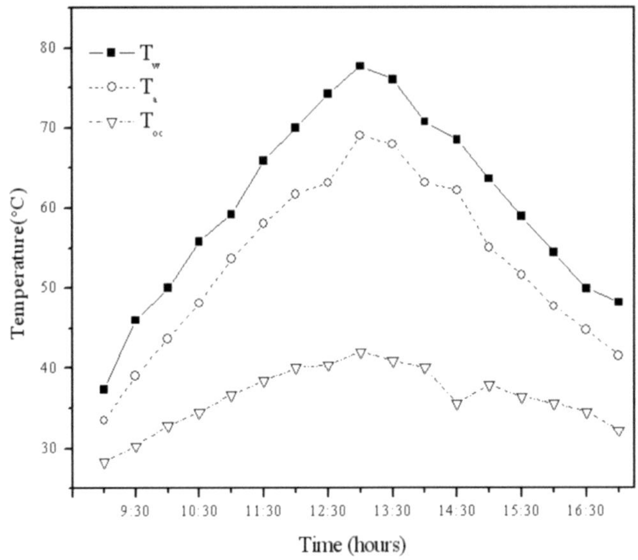

Figure 9. Hourly variation of temperature with cooling in pyramid solar still.

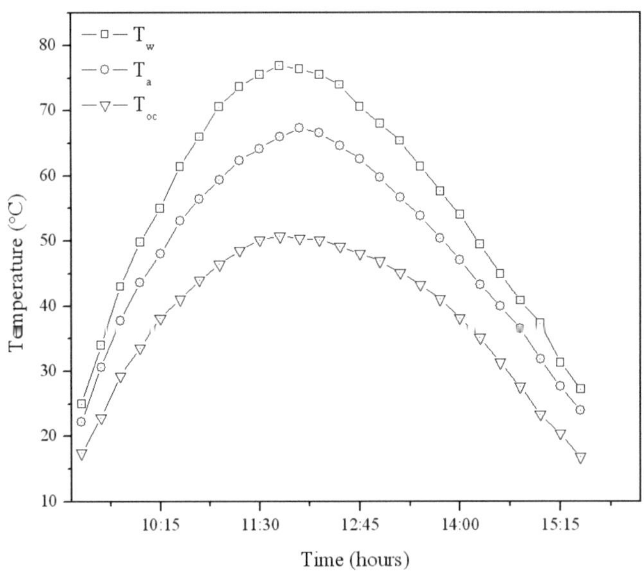

Figure 10. Hourly variation of temperature with cooling in CPC assembly.

http://dx.doi.org/10.5772/49995

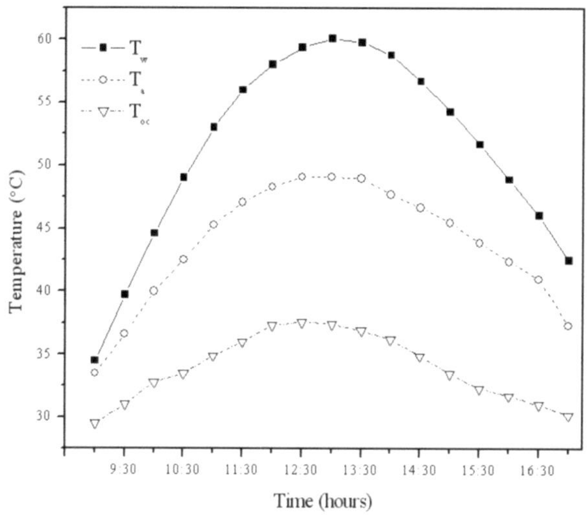

Figure 11. Hourly variation of temperatures for top cover closed pyramid solar still in CPC assembly.

Fig. 12 shows the variation of hourly production in the system. The distilled water collection is in the range of 160-625 ml for CPC-CCBTSS coupled with pyramid solar still without cooling and the total yield is collected as 6528 ml/m^2/day (Pyramid solar still 2818 ml/m^2/day and tubular solar still 3710 ml). Similarly, 160-650 ml for CPC-CCBTSS pyramid solar still with cooling and the total yield of 6928 ml/m^2/day (Pyramid solar still 3218 ml/m^2/day and tubular solar still 3710 ml) 40-470 ml for top cover closed pyramid solar still with CPC-CCBTSS and the total yield of 5243 ml/m^2/day (Pyramid solar still 1533 ml/m^2/day and tubular solar still 3710 ml are obtained from the above studies). The pyramid solar still operation is active mode when the heat extraction of water from CPC-CCBTSS is fed up into the pyramid solar still. Also the effective heat gained by the water in the pyramid solar still is estimated by top cover closed mode of operation. The yield rate is more than that of pyramid solar still alone. The productivity of the pyramid solar still is 2500 ml/m^2/day only. But it's yield rate is improved by coupling system with concentrator. A further increasing in yield rate is also observed from the extraction. In conventional solar still, the evaporation takes place after one hour from the beginning of experiment. This draw back has improved here. The warm-up period is reduced and supports for quick evaporation. The initial water temperature in the pyramid solar still is high as 55°C due to the heat extraction from the CPC-CCBTSS. Additionally, the heat extraction of water in the pyramid solar still temperature is further increased by the incoming solar radiation. So the water temperature is increased within a short interval of time. The sudden rise in water temperature is induced the evaporative heat transfer in the still. So the distilled yield increases more than conventional solar still. A further increase in yield rate is also observed for the cooling over the pyramid solar still under same mode of operation. Thus this result conformed that the assistance of concentrator certainly increased the yield rate of distilled water.

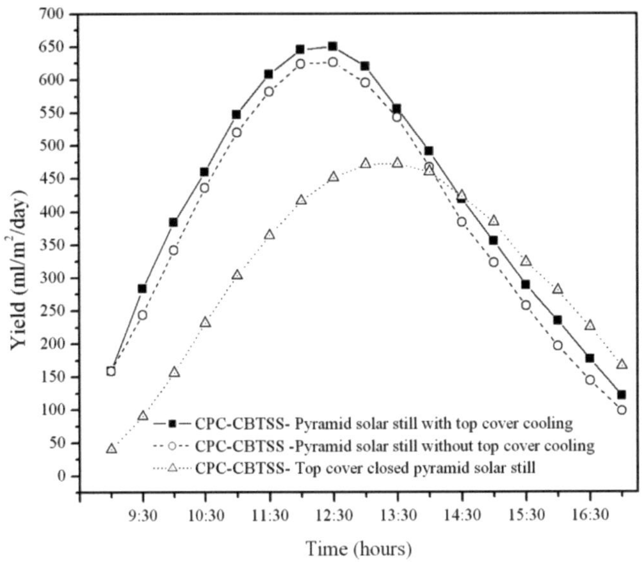

Figure 12. Hourly variation of yield with respect to time.

The temperature difference between the water in the basin and the glass cover are increased in the early hours from the beginning of the experiment due to the heat extraction of water from the CPC-CCBTSS which is directly coupled with pyramid solar still. The sum of yield has been taken into the calculation part. Also top cover cooling water from the CPC-CCBTSS is improved the operation of the solar still. So the energy loss of heat during cooling in the CPC-CCBTSS is further utilized by the pyramid solar still and increased the overall system efficiency. The overall efficiency of the system is calculated as 17.01% for without cooling and 21.14% for with cooling.

5. Conclusion

The temperature of the saline water in the basin can also be increased through the addition of external heating. These effects may be created by integrating with solar concentrator, flat plate collector. These type of behavior is studied by using of a pyramid solar still directly coupled with compound parabolic concentrator – tubular solar still in this work. It is completely different than that of other substitution effect to increase the temperature of basin water. CPC acts as a heat collecting unit and produces the distillate yield through circular basin cooling water from tubular cover serves as a further evaporation in the pyramid solar still. It can extract not only solar radiation incident on the saline water but also the other sources of heat loss. This type of utility of heat loss as a additional source will be suitable only at high temperatures. These results showed that the maximum output extracted from the proposed system as 6928 ml/m^2/day for with cooling.

Author details

T. Arunkumar[1], K. Vinothkumar[2], Amimul Ahsan[3], R. Jayaprakash[1*] and Sanjay Kumar[4]

*Address all correspondence to: jprakash_jpr@rediffmail.com

1 Solar Energy Laboratory, Department of Physics, Sri Ramakrishna Mission Vidyalaya College of Arts and Science, Coimbatore-641020, Tamilnadu, India

2 Research and Development (Renewable Energy), NSP Green Energy Technologies Pvt. Ltd. Chennai 603 209, Tamilnadu, India

3 Department of Civil Engineering, Faculty of Engineering, (Green Engineering and Sustainable Technology Lab, Institute of Advanced Technology), University Putra Malaysia, 43400 UPM Serdang, Selangor, Malaysia

4 Centre for Renewable Energy and Environmental Research, P.O. Box-5, Muzaffarpur-842001, Bihar, India

References

[1] Vinothkumar, K., & Kasturibai, R. (2008). Performance study on solar still with enhanced condensation. *Desalination*, 230, 51-61.

[2] Nijmeh, S., Odeh, S., & Akash, B. (2005). Experimental and theoretical study of a single-basin solar sill in Jordan. *Int Commun Heat Mass*, 32, 565-572.

[3] Tchinda, Réné., Kaptouom, Ernest., & Njomo, Donatien. (2000). Heat and mass transfer processes in a solar still with an indirect evaporator-condenser. *Energ Convers Manage*, 41, 93-107.

[4] Singh, A. K., & Tiwari, G. N. (1993). Thermal Evaluation of Regenerative Active Solar Distillation under Thermosyphon Mode. *Energ Convers Manage*, 34, 697-706.

[5] Chaouchi, Bachir., Zrelli, Adel., & Gabsi, Slimane. (2007). Desalination of brackish water by means of a parabolic solar concentrator. *Desalination*, 217, 118-126.

[6] Sinha, Kumar Sanjay. (1996). Transient model and comparative study of concentrator coupled regenerative solar still in forced circulation mode. *Energ Convers Manage*, 37, 629-636.

[7] Tiwari, G. N., & Sinha, S. (1993). Parametric study of active regenerative solar still. *Energ Convers Manage*, 34(3), 209-218.

[8] Murase, Kazuo., Tobata, Hiroshi., Ishikawa, Masayuki., & Toyama, Shigeki. (2006). Experimental and numerical analysis of a tube-type networked solar still for desert technology. *Desalination*, 190, 137-146.

[9] Tiwari, G. N., & Kumar, Ashok. (1988). Nocturnal water production by tubular solar stills using waste heat to preheat brine. *Desalination*, 69, 309-18.

[10] Ashan, Amimul., & Fikihara, Teruyuki. (2010). Mass and heat transfer of tubular solar still. *Sol Energy*, 841, 147-1156.

[11] Ashan, Amimul., Islam, K. M. S., Fukuhara, T., & Ghazali, A. H. (2010). Experimental study on evaporation, condensation and production rate of a new tubular solar still. *Desalination*, 260, 172-179.

[12] Amimul, Ashan., & Fukuhara, T. (2008). Condensation mass transfer in unsaturated humid air inside tubular solar still. *Annual Journal of Hydraulic Engineering, Japan of Civil Engineers*, 53, 97-102.

Section 5

Water Quality

Chapter 10

Water in Algerian Sahara: Environmental and Health impact

Khaled Sekkoum, Mohamed Fouzi Talhi,
Abdelkrim Cheriti, Younes Bourmita,
Nasser Belboukhari, Nouredine Boulenouar and
Safia Taleb

Additional information is available at the end of the chapter

http://dx.doi.org/10.5772/50319

1. Introduction

The supply of drinking water has become difficult in many countries, thus, access to safe drinking water, is expected to become in the the world as fundamental economic and social rights and unfortunately this is not currently the case. Whether its origin, superfecial or groundwater [Cheriti et al, 2011; Cheriti et al, 2009; Talhi et al, 2010], water used for human consumption are rarely consumables unchanged. It is often necessary to treat them more or less sophisticaly, or simply by disinfection in the case of groundwater. The reserves of groundwater in Algeria are estimated to 6.8 billion m^3. However, these groundwater are at significant depths and are characterized more by a strong mineralization, on the other hand, due to the particularity and specific climate of Algeria, the rivers dry frequently [ABHS, 2009]. As a rule, waters are subdivided into categories depending on a level of their mineralization or their rigidity. There are also other approaches to classification of water of various sources, for example, taking into account simultaneously its mineralization, rigidity and the contents of organic impurity. The boundary values for division of water into categories are sufficiently conventional and they differ in various sources of information [Djidel et al, 2010].

Better quality of water described by its physical, chemical and biological characteristics [Manjare et al, 2010]. The provision of good quality household drinking water is often regarded as an important means of improving health [Sanjana et al, 2011]. According to World Health Organization (WHO), there were estimated 4 billion cases of diarrhea and 2.2 million deaths annually [WHO, 2008].

The supply of clean water is limited by a lack of infrastructure, capacity and financial resources [Schafer et al, 2010]. Waterborne diseases caused by pathogens, long term exposure to chemicals such as fluoride. So, the quality of drinking water is becoming a serious public health issue for the past few years. The quality of water for drinking has deteriorated because of inefficient management of the piped water distribution system. The contamination of water with fecal material, domestic and industrial waste may result in an increased risk of disease transmission to individuals who use those waters [Radha Krishnan et al, 2007].

In another hand, in Algerian sahara, the drinking water supply is provided exclusively by groundwater from the aquifers of the Terminal Complex and Continental Intercalary (Albian water), characterized by high level of fluor, which its excessive consumption becomes toxic and constituting a public health problem especially in dental health [Bahloul et al, 2011].

The aim of the chapter focuses on the determination of water quality and assessing the possibility of wells in the Algerian sahara as an alternative source of drinking water and for domestic purposes. Characteristics of water are presented from different Saharan region. Analysis of the data shows that the general mineralization of water from the studied boring greatly exceeds the acceptable standards. Such water can be used only after demineralization. The fluoride content of drinking water of some Saharan region was measured, and health impact is discussed.

2. Characteristics of the Algerian Sahara

The Algerian Sahara is one of the hottest and driest in the world, covers an area of more than two million square kilometers and extends from the Saharan Atlas montains to the Malian, Nigerien, and Libyan borders. Its couvred a distance of over two thousand kilometers (north-south). This vast territory is formed by nine Wilayates (districts) with a population estimated at three million and a half inhabitants. The vast majority of this population is generally concentrated in the chief places of wilaya, some of which exceed 150,000 ca. The vast majority of this vast territory is occupied by large bodies represented by *regs*, *erg*, and *saline lakes*, which are unsuitable areas for agriculture. The Sahara is characterized by scarce rainfall and very irregular between 200 mm and 12 mm in the north to the south, high temperatures can exceed 45°C, accusing them of significant temperature fluctuations and also by low relative humidity of the air. Winds are relatively common and their speed is important from April to July, resulting in this period and the sirocco sandstorms, responsible for the formation and movement of dunes. The evaporation pan measured Colorado ranges from 2500 mm in the region of the Saharan Atlas and more than 4500 mm in the South (Adrar) [Djellouli-Tabet, 2010].

The Algerian Sahara is divided into four natural regions [ABHS, 2009]:

Chott Melghig: The hydrographic basin represented by Melghig Chott is one of the great watersheds of Algeria, it covers an area of 68,750 km^2, and it is distinguished by a large river of Oued Djedi. This river has many temporary tributaries that drain large areas and whose violent floods are sometimes devastating. Surges over the water and the deviations are numer-

http://dx.doi.org/10.5772/50319

ous, representing a non-negligible contribution water in irrigation perimeters and comes in extra-scooping into groundwater. Average annual rainfall in the basin varies between 200 and 300 mm / year.

Northern Sahara: This basin covers almost than 600,000 km² and is mainly distinguished by two important aquifers, the Continental Intercalary and the Terminal Complex [ABHS, 2009]. Groundwater provides drinking water supply and irrigation in this region. Rainfall in this region are very weak and random, varies from 200 mm / year in the north of the Saharan Atlas to the south 25mm/y.

Saoura region: Located in the South West of Algeria and covers an area of 320,273 km². It is limited by El Bayadh in the North, Mauritania in the South, Adrar in East and Morocco in West. The water surface potentials are very important (Djorf Torba Dam) and the groundwater quantity product could not solve the problem of water shortage for drinking water supply and irrigation. Indeed, all these water resources are conditioned by the contribution of rainfall, which is irregular and random [INC, 1983]. The rainfall in the region, may reach 200 mm in the north and decreases to 70 mm in the south of the region (Table 1 and 2) [AMS, 2009].

Tassili Nedjjar: This region is represented by the wilaya of Tamanrasset, which is characterized by a vast territory and a very low level of population regroupements. It covers an area of 556,100 km² and is bounded on the north by Ouargla and Ghardaia districts, to the east by Illizi, to the west by Adrar and in the south by Mali and Niger. The groundwater resources are very limited and are located mainly in the lap of the infero-flow of Oueds.

The Sahara region, is one of the vulnerable regions to climate change impacts. Climate change could have negative impacts on several socioeconomic sectors of the region like water resources.

Month	00h00	03h00	06h00	09h00	12h00	15h00	18h00	21h00
Jan.	6.7	5.5	4.6	7.6	12.9	14.6	12.3	8.8
Feb.	9.7	8.2	6.4	10.5	16.0	17.9	16.1	12.3
Mar.	14.4	12.5	11.4	15.9	20.2	21.6	19.9	16.2
Apr.	16.0	14.0	12.8	18.3	22.1	24.0	23.1	19.3
May	23.1	20.7	19.3	25.7	29.3	31.3	30.7	26.3
Jun.	27.0	25.7	24.3	31.0	34.6	36.1	35.3	30.8
Jul.	32.8	30.2	29.0	35.8	39.3	40.7	39.8	35.5
Aug.	32.1	30.0	28.3	34.3	38.4	40.1	38.5	34.6
Sept.	22.8	21.4	20.4	25.0	28.5	29.7	28.0	24.6
Oct.	20.2	18.2	16.6	23.5	28.1	29.6	27.3	22.9
Nov.	13.4	11.5	10.3	15.9	22.2	23.4	20.0	15.7
Dec.	10.3	8.8	7.4	11.3	18.0	20.3	16.5	12.7

Table 1. Monthly temperature rang in Algerian Sahara (1/10 C.) [AMS, 2009]

The first signs of changes already appear in this region through both the temperatures and the precipitation evolutions. Temperatures have increased by 1 to 2°C during the twentieth century. So, an important part of water resources in the Sahara has as origin the precipitation. Any changes in precipitation characteristics may affect water resources of this region already under water scarcity conditions [ANRH, 2001; OSS, 2001].

Month	00h00	03h00	06h00	09h00	12h00	15h00	18h00	21h00
Jan.	73.5	77.9	81.4	72.2	46.7	40.5	48.4	63.6
Feb.	64.4	72.3	78.0	65.6	42.5	34.9	39.2	53.9
Mar.	58.5	64.0	68.6	56.5	41.5	37.1	40.4	52.7
Apr.	52.4	56.7	60.9	44.8	33.3	28.1	29.4	39.5
May	39.5	44.8	49.9	36.1	28.9	24.0	23.0	29.4
Jun.	37.8	40.4	45.8	34.0	25.5	20.6	20.1	25.6
Jul.	23.1	26.2	29.3	23.7	18.7	15.4	14.3	18.9
Aug.	21.9	27.1	30.0	24.5	19.1	16.0	16.1	19.5
Sept.	62.0	66.2	69.6	53.7	41.9	38.1	41.4	53.6
Oct.	51.2	57.1	63.4	47.5	35.9	30.9	32.6	43.0
Nov.	51.6	57.3	62.1	52.8	38.4	33.6	37.2	46.3
Dec.	50.4	55.4	60.8	57.6	41.5	33.9	37.0	44.9

Table 2. Monthly relative Humidity rang in Algerian Sahara (%.) [AMS, 2009]

3. Water supply in Sahara Region

3.1. Potential water of resources in Algerian Sahara

In Algerian Sahara, the water has a vital character, because the climatic and hydrological contexts are extremely fragile. The spatial and temporal irregularity of the water availability, the impact of the droughts and flooding and the pressure of the demand of water are in continual increasing facing limited resources.

The potential water of resources in Algeria is of 17 Billion of m³ (surface water 10 Billion of m³, underground water 6.8 Billion of m³ mainly in the Sahara). According to Bouguerra [2001], the potential of the surface water resources in the north of Algeria, estimated at 13,500 hm³ per year in 1979, was reevaluated at 12,410 hm³ per year in 1986 and is more currently at only 9,700 hm³ per year. The resource is clearly declining, taking into account the dry conditions that have prevailed for the last three decades on all the basin slopes of northern Algeria as testified by the actual state of dams. The underground aquifers situated to the north of Algeria are exploited to 90%, with 1.9 Billion of m³ per year. Some aquifers are be-

http://dx.doi.org/10.5772/50319

coming overexploited and in the Sahara region the extracted volume is valued to 1.7 Billion of m^3. The most exploited water tables are less than 50 m deep, where they are easier and less expensive to reach. A growing number of wells tap groundwater between 100 m and more than 600 m deep. Various traditional and modern means are used to access these wells, including pulley and power-driven pumps [Meddi, 2006; Mutin, 2000].

The aquiferous system of the north Sahara, extending 1 million km^2, is shared by Algeria, Tunisia, and Libya. The groundwater tables are fed by the winter rains and sometimes by infiltration from the Oueds (Figure 1 and 2). Algeria, Tunisia, and Libya have launched efforts to coordinate the management of these water resources. The Aquiferous System of the Septentrional Sahara (SASS) is a program initiated by the international organization Observatory of the Sahara and the Sahel (OSS) to develop dialogue between the three countries.

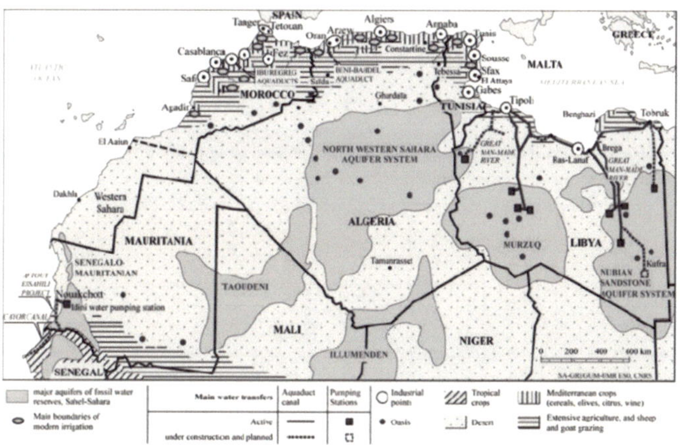

Figure 1. Groundwater resources and main water transfers in North Africa [Rekacewicz, 2006].

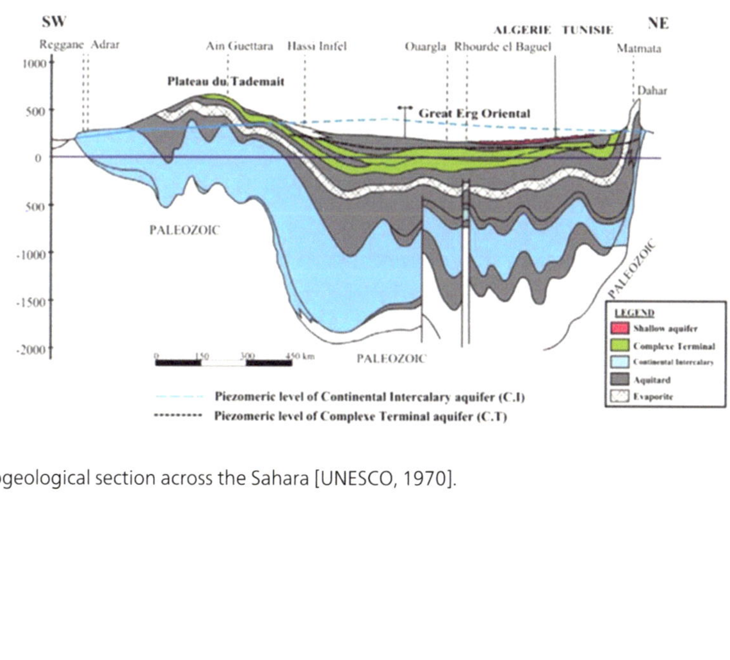

Figure 2. Hydrogeological section across the Sahara [UNESCO, 1970].

The number of pumping stations has multiplied between 1970 and 2000, to date, more than 7,000 water points exist in the countries. In 2002, a SASS report noted that "the simple continuation of the current intensity of pumping can constitute a serious danger." The volume of water pumped annually has increased by 525% in 50 years, from 0.4 billion m^3 in 1950 to 0.6 billion m^3 in 1970 and to 2.5 billion m^3 in 2000. The fact hat these resources are nonrenewable makes them even more vulnerable in the long term [Latreche, 2005; OSS, 2001].

Figure 3. Impact of flooding – El Bayadh City on October 01, 2011. a) Satellite images, Algerian Alsat-2A [ASAL, 2011], b), c) and d) Destruction of the Historical Mahboula Bridge.

The rainy season may occur in autumn, winter, or spring according to geographical position [Djellouli-Tabet, 2010; Le Houerou, 1984]. More than 120 million hectares (ha) of arid land is threatened by the processes of desertification, according to the United Nations Convention to Combat Desertification (UNCCD), with 445 million ha already considered desertified. Natural water resources are limited, and the spatial distribution and management of these different resources varies considerably depending on locality. The recorded rainfall amounts show that the average rainfall has decreased in recent years. In the last decade, this deficit in Algeria was more than 20% for the western area, 13% for the central part of the country, and 12% for the east [Bouguerra, 2001; Ould Amara, 2000].

Water scarcity in the Sahara may appear paradoxical given the exceptional weather events that led to recent floods in several areas that can be as well very prejudicial for the public or private infrastructures that for agriculture, and to cause numerous victims among the population.

http://dx.doi.org/10.5772/50319

The catastrophic floodings like the Timimoun in 2000, Ghardaia in 2008, Bechar in 2008, and recently El Bayadh on October 2011 (Figure 3), are causing the destruction of several infrastructures and the historical Mahboula Bridge.

3.2. Water Collection and Distribution

3.2.1. *Dams for the storage of surface water*

The storage of water during the humid years in order to use it during the dry years imposes itself in Algeria. The dams of bigger sizes should be encouraged to take into account the impact of the climatic changes. Thus, the Dams National Agency (ANB) is using more than 50 dams with a capacity of 5.1 billion m^3 and other dams are part of a project within the special framework emergency program to reinforce the drinking water supply in large cities.

The Algerian Sahara has a significant potential dams like: *El Gherza* (Sidi Okbba, Biskra), *Djorf torba* (Bechar) and *Larouia* (Brizina, El Bayadh) (Figure 4).

Figure 4. Dams in the Sahara, a) *Djorf torba*- Bechar- b) *Larouia*-Brezina.

3.2.2. *Reuse of the Foggara*

The need for people to overcome the challenges of an arid climate to meet water demand is nothing new. Traditional irrigation technique developed since ancient times in the region of the *Touat*, *Gourara* and *Tidikelt*, for capture and water supply to the aquifer through a system of draining galleries similar to qanats in Iran, khettaras in Morocco. These systems, called foggaras in Algerian sahara (Figure 5), constitute a remarkable water management and delivery network that has enabled people to live in arid environments. The system collects groundwater and carries it through small tunnels to irrigate gardens, allowing gravity irrigation. [Ahmadi et al, 2010; El Faiz et al, 2010; Senoussi et al, 2011].

Many official organizations as NOF (National Observatory of Foggara) recommend rehabilitation of this system by helping, repair, and monitor foggara systems in the oases of the Algerian Sahara to promote conservation and sustainable development of these lands, stabilize the populations of the oases, and reduce poverty and desertification.

Figure 5. Foggara system in Adrar - Algeria.

4. Ethopharmacological uses of water in Sahara

Water is an indispensable element in the lives of living beings and especially to that of man. Natural heritage, water indelibly marks people's identity, it became a factor in the development and archeology proves that, to 3800 BC, it is the progress of irrigation that allowed the rapid demographic growth observed in the Middle East. Culture, including religion, clearly influences how people perceive and manage a resource such as water. The water culture was very different events that have evolved over the ages by taking multiple expressions. It helped to disseminate techniques, behaviors, tastes refinement [Bouguerra, 2003; Cheriti et al, 2010].

4.1. Water: Practical and Traditional Knowledge

Ariha "Jericho", the oldest city in the world, founded in 8000 BC in the Judean Desert, owes its existence to the freshwater springs that form small natural lakes near the Dead Sea. Hammurabi (Babylon), dug canals and water rights codified in 1730 BC. Well water has become a key development issue and archeology proves that around 3800 BC, it is the improved irrigation techniques that have enabled the rapid demographic growth observed in the Middle East.

Local knowledge is a valuable resource that can contribute to improved development and are the basis for decision making in the areas of food security, human health, animal health, education and natural resource management. Its well knows that water is a preferred instrument of human gathering in traditional cultures, while a sink or source can gather a tribe nomadic or sedentary [Bouguerra, 2003; Faruqui et al, 2001; Ansari, 1994].

It is known that water in addition to its importance as a vital element in the life of living beings and especially to that of man, is a natural heritage, it marks indelibly the identity of people. Moreover, as water is also found in symbolic practices and traditional knowledge, it affects our environment and our daily realities. The water culture was very different events that have evolved over the ages by taking multiple expressions. It helped to disseminate techniques, behaviors, tastes of refinement

It is observed clearly from the practice of special chemical of water, so simple molecule composed of atoms them essential to life, Hydrogen and Oxygen the particularity to pass a physical state to another: solid (ice), liquid (water) and gas (steam). Similarly it is through water that passes from one state to another: Dirt / Clean, Fatigue / Relaxation, Disease / Health and Life / Death [Cheriti et al, 2010].

4.2. Water ethnopharmacological practice (for example the region of El Bayadh)

Water as a symbol is also found in the practices and traditional knowledge; it affects our environment and our daily realities. The use of water is crucial in public health. Lack of access to water and a healthy environment is one of the first direct or indirect causes of death and disease in the world. Annually are 250 million people who suffer from diseases caused by water undrinkable. In developing countries, 80% of cases of diseases identified are rooted in poor water quality. Algeria is its strategic location and history, its large area, its diverse climate, its flora varies, has a source of materia medica and a rich and abundant traditional skills important.

We present the results in our statement of our surveys in the southwest of the Algerian instead of water in the local ethnopharmacology [Cheriti et al; 2010]. We conducted a study in El Bayadh district on the importance of water in local traditional medicine, led us to the following points:

The traditional herbal preparation use water as essential solvent for extracting, compared to fat or vinegar:

- Water (decoction, infusion, maceration...): 73%
- Material Fat: 14%
- Vinegar / Alcohol: 6%
- Other (direct saliva....): 7%

The same survey conducted in six localities in the El Bayadh district, lets us conclude that water sources is especially recommended in traditional medicine, in particular for the treatment of dermatological diseases, urinary and ophthalmic disorder (Figure 6 and 7). The most remarkable thing that the therapeutic effects of water from Ain El Mahboula remain etched in the memories of local people despite the disappearance of the source for over twenty years. Because a generation aged 15 to 25 years does not know the source only through orality relatives from old.

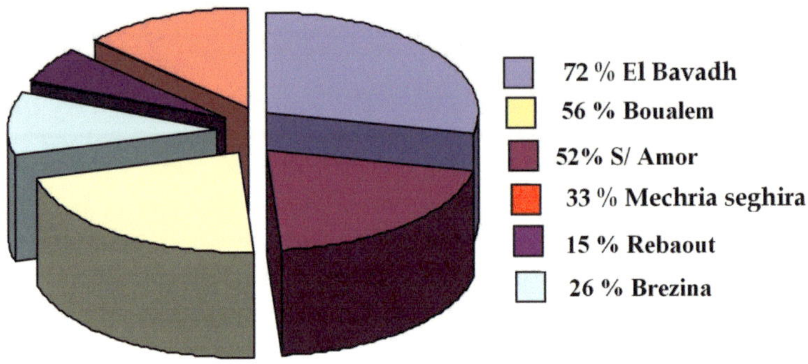

Figure 6. Frequency of traditional use of water in localities of El Bayadh district.

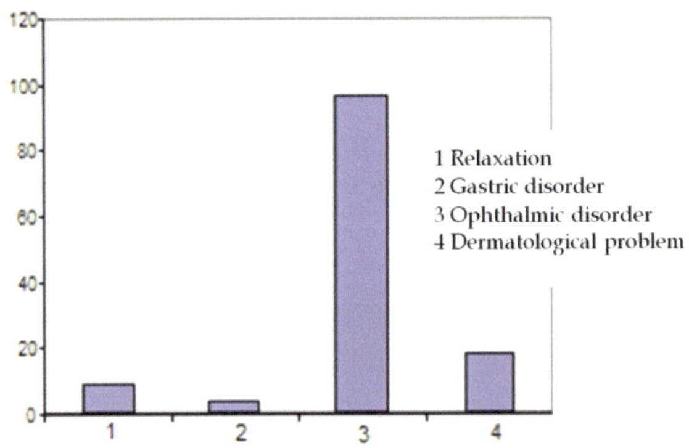

Figure 7. Traditional treatment of diseases by traditional El Mahboula spring water source.

Finally, we consider water as a heritage both natural and cultural. In southwest Algeria there is a traditional knowledge management and the involvement of water sources especially for the treatment of various pathologies. It is necessary to consider the socio-cultural interactions with other components of the natural environment for the management of water resources and enhance the traditional expertise and local medicinal heritage.

5. Analysis of the Saharan water

Chemical analysis of water from different Sahara localities (Table 3), shows that the general mineralization of water from the studied boring greatly exceeds the acceptable standards. Such water can be used only after demineralization.

The anions (Cl^-, $SO_4{}^{2-}$) show that they are not in the standards of water potability in both Tindouf, Bechar and Ouargla localities, which made water unsafe to drink. Chloride toxicity has been observed in humans, as in the case of impaired metabolism of sodium chloride (NaCl), for example, in congestive heart failure [WHO, 1996].

http://dx.doi.org/10.5772/50319

Location	pH	Cond (s/cm)	Rs (mg/l)	Ca^{++} (mg/l)	Mg^{++} (mg/l)	Na^{+} (mg/l)	K^{+} (mg/l)
Tindouf	7.36	3.66	2270	202	114	416	11
Bechar	7.29	4.09	2536	161	142	460	16
Adrar	7.28	2.08	1310	111	101	160	6
Naama	7.27	2.22	1380	120	111	190	9
El Bayadh	7.1	2.2	1250	91	109	160	5.4
Ghardaia	7.2	2.33	1226	21	149	145	13
Ouargla	8.23	4.27	2916	193	287	480	29
Illizi	7.73	4.58	4982	200	252	604	26
OMS guideline value			2000	100	250	200	-

Location	Cl-(mg/l)	SO_4^{2-} (mg/l)	HCO_3-(mg/l)	NO_3- (mg/l)
Tindouf	662.5	655	210	92
Bechar	730	750	207	89
Adrar	245	380	311	64
Naama	305	395	268	73
El Bayadh	268	325	250	54
Ghardaia	265	400	128	10
Ouargla	895	725	102	14
Illizi	265	2300	128	7
OMS guideline value	250	400	-	44

Table 3. Parameters of some waters from Algerian Sahara.

Healthy people can tolerate the consumption of large quantities of chloride provided there is a concomitant intake of fresh water [Djidel et al., 2010], little is known about the effect of prolonged ingestion of large amount of chloride in diet [Kesteloot et al, 1988]. Experimental studies show that hypertension associated with ingestion of sodium chloride seem to be related to the sodium rather than chloride ion [Haijar et al, 2001]. However, high level of sulfate in water can provide dehydration and diarrhea and children are often more sensitive to sulfate than adults [NCEH, 1999].

The cations (Na^+, K^+) show that they are not in the standards of potability of water, which made water very salty and hard. Calcium concentration in this water is higher than OMS guidelines value. The abuse of calcium ingestion without medical advice can lead to the development of blood clots, kidney problem such as urolithiasis and potassium accumulation can cause a disturbance of heart beats [Djellouli et al, 2005; Sekkoum et al, 2012a]. When the concentration of Ca and Mg decrease, the concentration of sodium and SAR index become more important [Eriksen et al, 1990]. This will cause an alkalizing effect and increase the pH. Therefore, when a water test indicates a high pH, this may be a sign of high content of carbonate and bicarbonate ions [Djellouli et al, 2005].

The carbonate and bicarbonate combined with calcium or magnesium will precipitate as calcium carbonate ($CaCO_3$) or magnesium carbonate ($MgCO_3$) in dry condition (the Sahara shows a permanent drought), so, the classification of these waters shows very high salinity water. [Kumar et al, 2010].

In another hand we conducted a study of water quality and assessing the possibility of wells in the Bechar district (South-West Algeria) as an alternative source of drinking water and for domestic purposes.

The artesian well of Mougheul, with depth of 185 meters is one of potential sources of water supply of district, and allows to provide initial water productivity up to 8 l/s. Requirements to the water for potable needs in district is presented according to specifications of the EU and the World Health Organization [WHO, 2004].

For definition of the opportunity to use the well as a source of water for potable and domestic needs, samples have been taken and the basic physical and chemical parameters of water are determined according to standard techniques. In table 4 are presented the analysis results of water quality of the Mougheul well in comparison with the requirements regulating quality of potable water within the framework of the international standards.

The analysis of data (table 4) shows, that the general mineralization of water of Mougheul well considerably surpasses allowable requirements of all above-mentioned standards.

As a rule, waters are subdivided into categories depending on a level of their mineralization or their rigidity. There are also other approaches to classification of water of various sources, for example, taking into account simultaneously its mineralization, rigidity and the contents of organic impurity [Gousseva et al, 2000]. The boundary values for division of water into categories are sufficiently conventional and they differ in various sources of information. The type of water and the contents of the basic impurity in it allow to choose correctly a method of its conditioning, and also to pick up the most effective materials and the equipment for water preparation.

According to the above mentioned data (Table 4), water of Mougheul well in the Béchar district can be classified in the category of salty underground waters with high hardness. The most suitable method of conditioning of such water to have a quality up to a level of the required norms for potable and domestic to water is the technology of barometric membrane or the combined technology of barometric membrane with ionic interchange method of water treatment. The development of basic alternative technological schemes of water-preparation and the evaluation of economic parameters of these schemes will allow to choose the most rational and economic scheme of water conditioning to have its quality up to a level for potable water. The development of rational technology of water conditioning received from the artesian well of Mougheul in the wilaya of Béchar will allow to receive drinking water quality and to minimize economic expenses for process of water treatment. The introduction of such installations will allow keeping resources of dams and other existing sources of water supply for more remote regions of the country and, that is important for needs of agriculture.

Parameters	Value of parameter		
	Well of Mougheul	Normative requirements	
		WHO	European Union
pH	7,4	6,5-8,5	6,5-8,5
Overall hardness, mg-equiv./l	53,2	-	-.
Calcium, mg-equiv./l	40,6	-	100
Magnesium, mg-equiv./l	12,6	-	50
Overall alkalinity, mg-equiv./l	2,64	-	0,5
Potassium, mg/l	8,04	-	12
Sodium, mg/l	1876	200	200
Overall iron, mg/l	2,54	0,3	0,2
Overall manganese, mg/l	<0,01	0,1	0,05
Nitrates, mg/l	15,6	50	50
Sulfates, mg/l	2496	250	250
Chlorides, mg/l	2851	250	250
Bicarbonates, mg/l	161	-	-
Silicates, converted in SiO_2, mg/l	23,4	-	-
Fluorides, mg/l	1,0	1,5	1,5
Oxidability, mgO_2/l	3,2	-	5,0
Overall mineralization, mg/l	8418	1000	1500

Table 4. Parameters of water from Mougheul well.

6. Fluoride content of drinking Sahara water and health impact

Fluoride is an essential element to prevent carious dental [Sohn et al, 2007]. Incorporated into the teeth, fluoride decreases the solubility of enamel in acid medium which is consists mainly of hydroxyapatite and favorize the remineralization of initial carious lesions of enamel [Singh et al, 2003]. Water is the main source of fluoride ions [Emmanuel et al, 2002; Featherstone, 2000].

In southern Algeria, the drinking water is characterized by high level of fluor. However, excessive consumption of this oligo-element becomes toxic. Thus, in 2001, endemic areas of fluorosis were detected in Algerian sahara (El-Oued, Touggourt, Biskra, Timimoun, Ouargla and Ghardaïa), constituting a public health problem caused by the ingestion of an excess of fluoride. In this respect, may be these regions are not concerned by the program of oral-dental health in schools [NPOHS, 2006]. In southern areas, where temperatures are high, the daily intake of water becomes more important. The standards of the World Health Organization (WHO) set at 0.8 mg/L the maximum concentration of fluorine permissible for public distribution water in these warm regions [Sekkoum et al, 2012b; WHO, 2006; 2004].

The concentration levels of samples taken from different location in the Algerian Sahara ranged from very low concentration (0.4 mg/L.) to very high level (4.32 mg/L.). As expected, most source of high fluorides levels were found in public distribution waters from the Wilaya (district) of Biskra, Adrar, Ouargla (Figure 8).

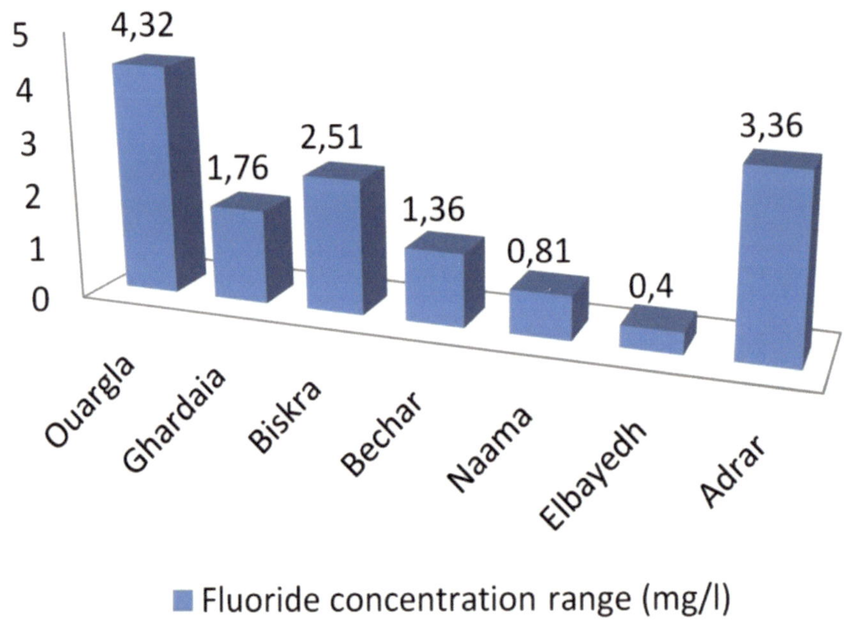

Figure 8. Fluoride levels in drinking water from South Algeria.

Taleb team [Bahloul et al, 2011] were studied the action of some Saharan waters containing different concentrations of fluoride as an inhibitor on dissolution of the hydroxyapatite. The efficacy of fluoride ions contained in the tested waters was evaluated according to the ratio of inhibition rate in presence of the drinking water and synthetic water. As indicated in Table 5, the inhibitory effects of Saharan waters are located within a range of 54.28 and 83.1%.

http://dx.doi.org/10.5772/50319

Water samples	VE (mL)	I %	I ref %	E = I/Iref %
Reference [F ⁻] = 0 mg/L	17.50	0	-	0
El bayadha (O. souf)	5.25	70	90.57	77.77
Tolga(Biskra)	4.80	72.5	90	80.55
Beldate amor (Tougourt)	3.90	77.71	89.54	86.78
El Guemar (O. Souf)	5.10	70.8	89.42	79.17
El Chott (O. Souf)	2.95	83.1	89.14	93.22
Hassi Messaoud	3.20	81.7	88	92.84
Oued Souf	4.10	76.57	88	87.01
Reganne	4.60	73.71	83.71	88.05
Adrar	5.30	69.71	84	82.98
Touggourt	8.0	54.28	85	63.85
Ksar Hirane (Laghouat)	4.80	72.57	77.71	93.38
Zelfana (Ghardaia)	6.80	61.14	61.71	99.07
Hassi R'Mel	7.80	55.43	65.14	85.09
El Golea	6.50	62.85	61.71	101.80

Table 5. Inhibitory effect of Saharan water on the dissolution of hydroxyapatite.

Moreover, Dissananyake [1991], showed that dental carious occurs in region where drinking water is less fluoridated, while it is absent in areas with fluorine rich water. Other studies have indicated that fluoridation of water is very important to maintain the bucco-dental health [Angelillo et al, 1999; Levy, 2003]. In contrary, according to the World Health Organization [WHO, 2006], the fluoride rich water causes a risk of dental fluorosis. Indeed, the amount of fluorine called "optimal dose of fluoride in drinking water" which decreases the prevalence of dental carious with the absence of a significant fluorosis varies between 0.7 and 1.2 mg/L [Emmanuel et al, 2002]. Finally, if we consider the influence of temperature, all public supply waters of the south Algeria are excessively fluoridated. On the other hand, the experimental approach, in vitro, shows the importance of fluoride of drinking water from southern Algeria in preventing dental carious. Indeed, the extremely high temperature of the South is a major factor contributing to the increase in demand for drinking water and, consequently, the increase in dental fluorosis. Therefore, to reduce this risk in this region, consumers need to correct their food habits not exceeded the needs of the body in fluorine. Thus, 0.05 to 1 mg of fluoride are considered as not toxic daily dose on the health of the adult population.

7. Conclusion

Demographic, social, and economic factors will determine the future demand and availability of water resources. Generally, the groundwater constitutes an important part of the hydraulic heritage of the Algerian Sahara. To assure their safeguard and their protection for the future generations and for the difficult moments, it is necessary to : Control the withdrawals of underground waters in order to protect them against the overexploitation; to reinforce the integrated management of water resources and water policy, generalize the economy of water for all users and sensitization of the users, the local actors, the local decision makers and the agents of authority to the constraints of water scarcity and to the risk of draining of underground waters.

In the context of the scarcity, the quality of the superficial and underground waters is threatened by numerous problems. It is important to undertake all to improve and to preserve the quality of water resources and the protection of the environment.

It is necessary to consider the socio-cultural interactions with other components of the natural environment for the management of water resources and enhance the traditional expertise and local knowledge. In another hand the study shows the importance of fluoride of drinking water from Algerian Sahara in preventing dental carious.

Author details

Khaled Sekkoum[1,2], Mohamed Fouzi Talhi[1], Abdelkrim Cheriti[1*], Younes Bourmita[1], Nasser Belboukhari[2], Nouredine Boulenouar[2] and Safia Taleb[3]

*Address all correspondence to: karimcheriti@yahoo.com

1 Phytochemistry & Organic Synthesis Laboratory, University of Bechar, Algeria

2 Bioactive Molecules & Chiral Separation Laboratory, University of Bechar, Algeria

3 Catalysis & Materials Laboratory, University D. Liabes Sidi Bel Abbès, Algeria

References

[1] ABHS (2009). The Sahara hydrographic basin agency, letter n 4.

[2] Ahmadi, H., Samani, A. N., & Malekian, A. (2010). The Qanat: A Living History in Iran, Chapter 8 in Water and Sustainability in Arid Regions, Ed Schneier-Madanes G. & Courel M.F., Editors, Springer Science, Heidelberg.

[3] AMS (2009). Algerian Metrological Station report.

[4] Angelillo, I. F., Torre, I., & Nobile, C. G. (1999). Caries and fluorosis prevalence in communities with different concentrations of fluoride in the water. *Caries Research*, 33(2), 114-122.

[5] ANRH, Agence Nationale des Ressources Hydrauliques. (2001). Banque de données pluviométriques et hydrologiques d'Algérie.

[6] Ansari, M. I. (1994). Islamic Perspectives on Sustainable Development. *American Journal of Islamic Social Science*, 11(3), 394-402.

[7] ASAL (Algerian Space Agency). (2011). http://www.asal.dz/inondations-el-bayadh.php.

[8] Bahloul, H., Djellouli, H. M., Taleb, S., Rezgane, Z., Cheriti, A., & Benghalem, A. (2011). In vitro inhibitory effect of drinking water from south Algeria on the dissolution of dental hydroxyapatite. *Med. J. Chem.*, 1(4), 192-199.

[9] Bouguerra, K. (2001). Potentialités en ressources en eau superficielle du nord du pays (Algérie). No 71/ANRH/DHYL, 10.

[10] Bouguerra, M. L. (2003). Symbolique et culture de l'eau, rapport N 5, Institut Veolia Environnement.

[11] Cheriti, A., Talhi, M. F., Belboukhari, N., & Taleb, S. (2011). Copper Ions Biosorption properties of Biomass derived from Algerian Sahara plants. , Chapter 16, in Expanding issues in Desalination, Edited by Robert Y. Ning, InTech Publishers, Austria.

[12] Cheriti, A., Talhi, M. F., Belboukhari, N., Taleb, S., & Roussel, C. (2009). Removal of copper from aqueous solution by Retama raetam Forssk. growing in Algerian Sahara. *Desalination and Water Treatment*, 10, 317-320.

[13] Cheriti, A., & Sekkoum, K. (2010). Ain Mahboula from El Bayadh, betwen heritage and memory(in arabic). *Al Athar Journal*, 6, 93-101.

[14] Djellouli, H. M., Taleb, S., Harrache-Chettouh, D., & Djaroud, S. (2005). Qualité physico-chimique des eaux de boisson du Sud Algérien: étude de l'excès en sels minéraux. *Cahiers Santé*, 15(2), 109-112.

[15] Djellouli-Tabet, Y. (2010). Common Scarcity, Diverse Responses in the Maghreb Region. , Chapter 6, in Water and Sustainability in Arid Regions, Ed Schneier-Madanes G. & Courel M.F., Editors, Springer Science, Heidelberg.

[16] Djidel, M., Bousnoubra-Kherici, H., & Nezli-E, I. (2010). The Minerality Impact of Deep Groundwater, in Desert Regions, on Human and the Environment. South east Algeria. *European Journal of Scientific Research*, 45(4), 540-551.

[17] Dissanayake, C. B. (1991). The fluoride problem in the groundwater of Sri Lanka-environmental management and health. *International Journal of Environmental Studies*, 38, 137-156.

[18] El Faiz, M., & Ruf, T. (2010). An Introduction to the Khettara in Morocco: Two Contrasting Cases. , Chapter 10, in Water and Sustainability in Arid Regions, Ed Schneier-Madanes G. & Courel M.F., Editors, Springer Science, Heidelberg.

[19] Emmanuel, E., Fanfan, P. N., Louis, R., & Michel, G. A. (2002). Determining the optimal fluorine dose in the drinking water of the South Center hydrological region of Haiti. *Cahiers Santé*, 12(2), 241-245.

[20] Eriksen, E. F., Hodgson, S. F., Eastell, R., Cedel, S. L., O'Fallon, W. M., & Riggs, B. L. (1990). Cancellous bone remodeling in type I (postmenopausal) osteoporosis: quantitative assessment of rates of formation, resorption, and bone loss at tissue and cellular levels. *J. Bone Miner Res.*, 5(4), 311-319.

[21] Faruqui, N. I., Biswas, A. K., & Bino, M. J. (2001). *Water management in Islam*, United Nations University Press.

[22] Featherstone, J. D. (2000). The science and practice of caries prevention. *J. Am. Dent. Assoc.*, 131(7), 887-899.

[23] Gousseva, T. V., Moltchanova, Ya. P., Zaika, E. A., Vinnitchenko, V. N., & Averotchkin, E. M. (2000). The hydrodynamic parameters of environment state. Reference documents, Edition Ekoline, Ukrania.

[24] Haijar, I. M., Grim, C. E., Gerge, V., & Kotchen, T. A. (2001). Impact of diet on blood pressure and age-related changes in blood pressure in the U.S. population. *Arch Intern Med.*, 161, 589-593.

[25] INC (1983). Atlas of Algeria Geography and Maps Technol.

[26] Kesteloot, H., & Joosens, J. V. (1988). Relationship of dietary sodium potassium, calcium, and magnesium with blood pressure. Belgian interuniversity Research on Nutrition and Health. *Hypertension*, 12, 594-599.

[27] Latreche, D. (2005). Connaissance et exploitation des ressources en eau partagées au Sahara septentrional (SASS/OSS). Colloque international sur les ressources en eau dans le Sahara. Conférence Ouargla, Algérie.

[28] Le Houerou, H. N. (1984). Rain use efficiency a unifying concept in arid land ecology. *Journal of Arid Environments*, 7, 213-247.

[29] Levy, S. M. (2003). An update on fluorides and fluorosis. *J. Can. Dent. Assoc.*, 69(5), 286-291.

[30] Manjare, S. A., Vhanalakar, S. A., & Muley, D. V. (2010). Analysis of water quality using physico-chemical parameters tamdalge tank in kolhapur district, maharashtra. *International Journal of Advanced Biotechnology and Research*, 1(2), 115-119.

[31] Meddi, M. (2006). Evolution des régimes pluviométriques dans les différentes stations du nord du Sahara septentrional. Avenir des zones sèches, Tunis: UNESCO, 10.

[32] Mutin, G. (2000). L'eau dans le monde arabe. Enjeux et conflits. Paris, Ellipses Edition.

[33] NCEH (National Center for Environmental Health) (1999). Health Effects from Exposure to High Levels of Sulfate in Drinking Water Study, National Center for Environmental Health.

[34] NPOHS (National Program for Oral Health in Schools) (2006). The interministerial circular of 27 March 2006.

[35] Kumar, N., & Sinha, D. K. (2010). Drinking water quality management through correlation studies among various physicochemical parameters: A case study. *International journal of environmental sciences*, 1(2), 253-259.

[36] OSS (Observatory of the Sahara and the Sahel) (2001). Les ressources en eau des pays de l'Observatoire du Sahara et du Sahel, évaluation, utilisation et gestion. 87.

[37] Ould Amara, A. (2000). La sécheresse en Algérie. ANRH, Bir Mourad Raïs, 5.

[38] Radha Krishnan, R., Dharmaraj, K., & Ranjitha Kumari, B. D. (2007). A comparative study on the physicochemical and bacterial analysis of drinking, borewell and sewage water in the three different places of Sivakasi. *Journal of Environmental Biology*, 28(1), 105-108.

[39] Rekacewicz, Ph. (2006). Visions cartographiques. *Le Monde diplomatique*, Paris, http://blog.mondediplo.net/-Visionscartographiques.

[40] Sanjana, C., Menka, B., & Sarita, O. (2011). Physiochemical Characterization of Some Drinking Water Samples of Jaipur City. *Water Research & Development*, 1(1), 53-55.

[41] Schafer, A. I., Rossiter, H. M. A., Owusu, P. A., Richards, B. S., & Awuahb, E. (2010). Developing Country Water Supplies : Physico-Chemical Water Quality in Ghana. *Desalination*, 251, 193-203.

[42] Sekkoum, K., Cheriti, A., Taleb, S., & Belboukhari, N. (2012a). FTIR spectroscopic study of human urinary stones from El Bayadh district (Algeria). *Arabian Journal of Chemistry*, Under Press.

[43] Sekkoum, K., Cheriti, A., & Taleb, S. (2012b). Fluoride analysis in water from El Bayadh district(in arabic). *Al Badr Review*, 4(2), 17-20.

[44] Senoussi, Bensania M., Moulaye, S., & Telli, N. (2011). la foggara: Un système hydraulique multiséculaire en déclin, Revue Des Bioressources, 1, 1.

[45] Singh, K. A., Spencer, A. J., & Armfield, J. M. (2003). Relative effects of pre and post-eruption water fluoride on caries experience of permanent first molars. *J. Public Health Dent.*, 63(1), 11-19.

[46] Sohn, W., Ismail, A. I., & Taichman, L. S. (2007). Caries Risk based fluoride supplementation for children. *Pediatre Dent.*, 29, 23-29.

[47] Talhi, M. F., Cheriti, A., Belboukhari, N., Roussel, C., & Taleb, S. (2009). Biosorption of Heavy metals from Aqueous Solutions by Retam reatem plant. *Desalination & Water treatment*, 10(1-3), 317.

[48] UNESCO (1970). Study of water resources of the northern Sahara, Eressëa Project, Final Report- Paris.

[49] WHO (World Health Organization) (1996). Guidelines for drinking-water quality Health criteria and other supporting information, 2nd ed. , World Health Organization, Geneva., 2

[50] WHO (World Health Organization) (2004). Guidelines for Drinking-Water Quality Recommendations. , 1, Third Edition, Geneva, 494.

[51] WHO (World Health Organization) (2006). Guidelines for Drinking-Water Quality Recommendations. , Geneva.

[52] WHO (World Health Organization) (2008). World Health Organization Report, Geneva.